SpringerBriefs in Statistics

More information about this series at http://www.springer.com/series/8921

Nina Golyandina · Anatoly Zhigljavsky

Singular Spectrum Analysis for Time Series

Second Edition

 Springer

Nina Golyandina
Faculty of Mathematics and Mechanics
St. Petersburg State University
St. Petersburg, Russia

Anatoly Zhigljavsky
School of Mathematics
Cardiff University
Cardiff, UK

ISSN 2191-544X   ISSN 2191-5458   (electronic)
SpringerBriefs in Statistics
ISBN 978-3-662-62435-7   ISBN 978-3-662-62436-4   (eBook)
https://doi.org/10.1007/978-3-662-62436-4

This Springer imprint is published by the registered company Springer-Verlag GmbH, DE part of Springer Nature.
The registered company address is: Heidelberger Platz 3, 14197 Berlin, Germany

Preface to the Second Edition

Singular spectrum analysis (SSA) is a technique of time series analysis and forecasting. It combines elements of classical time series analysis, multivariate statistics, multivariate geometry, dynamical systems and signal processing. SSA can be very useful in solving a variety of problems such as forecasting, imputation of missing values, decomposition of the original time series into a sum of a small number of interpretable components such as a slowly varying trend, oscillatory components and a 'structureless' noise. For applying the core SSA algorithms, neither a parametric model nor stationarity-type conditions have to be assumed. This makes some versions of SSA model-free, which enables SSA to have a very wide range of applicability. The rapidly increasing number of new applications of SSA is a consequence of new fundamental research on SSA and the recent progress in computing and software engineering, which made it possible to use SSA for very complicated tasks that were unthinkable 20 years ago, see Sect. 1.3.

The subject of time series analysis and forecasting is very well developed. Despite its undisputable usefulness, SSA has not yet reached the popularity it deserves. We see the following four main reasons why SSA is still not recognized by certain groups of statisticians. The first reason is tradition: SSA is not a classical statistical method and many statisticians are simply not aware of it. Second, mainstream statisticians often prefer model-based techniques where calculations are automatic and do not require the computer-analyst interaction. Third, in some instances SSA requires substantial computing power. Finally, SSA is sometimes too flexible (especially when analyzing multivariate time series), and therefore has too many options which are difficult to formalize.

The first edition of this book has been published in 2013 and has attracted considerable attention of specialists and those interested in the analysis of complicated time series data. We believe this is related to the special status of this book as the sole publication where the methodology of SSA is concisely but at the same time comprehensively explained. The book exhibits the huge potential of SSA and shows how to use SSA both safely and with maximum effect.

A mutual agreement between the publisher and the authors concerning the second edition of the book has been easily reached. A permission of volume increase by almost 15% has been granted. We have used this possibility to broaden the range of topics covered; we have also revised parts of the first edition of the book in view of recent publications on SSA and our deeper understanding of the subject. The two main additions to the previous version of the book are: (a) Sect. 1.2, where we discuss the place of SSA among other methods, and (b) Sects. 2.6 and 3.10 devoted to multivariate and multidimensional extensions of Basic SSA. Also, we have made significant changes in all sections of Chap. 1 as well as in Sects. 2.4.4, 2.5.2, 2.5.3 and 3.8.4. Finally, we have added Sect. 2.5.5, where we discuss the ways of dealing with outliers. All other sections of the present edition are very similar to the corresponding sections in the first edition; we just corrected a few typos, improved certain explanations and slightly modified some text to make it consistent with the new material.

Potential readers of the book are: (a) professional statisticians and econometricians; (b) specialists in any discipline where problems of time series analysis and forecasting occur; (c) specialists in signal processing and those needed to extract signals from noisy data; (d) Ph.D. students working on topics related to time series analysis; (e) students taking appropriate M.Sc. courses on applied time series analysis; (f) anyone interested in the interdisciplinarity of statistics and mathematics.

Acknowledgements The authors are very much indebted to Vladimir Nekrutkin, a coauthor of their first monograph on SSA. His contribution to the methodology and especially theory of SSA cannot be underestimated. We especially want to thank Anton Korobeynikov, who is the original author and the maintainer of the R-package Rssa with fast computer implementation of SSA and also our coauthor of the book devoted to the SSA-family methods and their implementation in Rssa. The recent development of SSA for long time series and multidimensional extensions would be impossible without effective implementation. The authors very much acknowledge many useful comments made by Jon Gillard. The authors are also grateful to former and current Ph.D. students and collaborators of Nina Golyandina: Konstantin Usevich, a specialist in algebraic approach to linear recurrence relations (2D-SSA), Theodore Alexandrov (automatic SSA), Andrey Pepelyshev (SSA for density estimation), Maxim Lomtev (SSA-ICA), Eugene Osipov and Marina Zhukova (missing data imputation), Alex Shlemov (SSA filtering, versions of SSA for improving separability). The help of Alex Shlemov in preparation of figures is very much appreciated.

St. Petersburg, Russia   Nina Golyandina
Cardiff, UK   Anatoly Zhigljavsky
August 2020

Contents

Chapter 1
Introduction

1.1 Overview of SSA Methodology and the Structure of the Book

The main body of the present book consists of Chaps. 2 and 3. In Chap. 2, SSA is normally considered as a model-free method of time series analysis. The applications of SSA dealt with in Chap. 3 require the use of models and hence SSA of Chap. 3 is model-based (Sect. 3.9 is an exception). As the main model, we use the assumption that the components of the original time series, which are extracted by SSA, satisfy (at least, locally) linear recurrence relations.

Let us briefly describe the structure of the book and at the same time make a short walk through the SSA methodology.

The algorithm of Basic SSA (*Sect. 2.1*). Basic SSA, the main version of SSA, consists of the following four steps, which we formulate here briefly and informally.

Step 1: Embedding. For a time series $\mathbb{X}_N = (x_1, \ldots, x_N)$ of length N and a fixed window length L ($1 < L < N$), we construct the L-lagged vectors $X_i = (x_i, \ldots, x_{i+L-1})^{\mathrm{T}}$, $i = 1, 2, \ldots, K$, where $K = N - L + 1$, and compose these vectors into a matrix $\mathbf{X}$ of size $L \times K$. The matrix $\mathbf{X}$ is called 'trajectory matrix'; the columns X_i of $\mathbf{X}$ are considered as vectors in the L-dimensional space R^L.

Step 2: Singular value decomposition (SVD). The eigendecomposition of the matrix $\mathbf{X}\mathbf{X}^{\mathrm{T}}$ (equivalently, the SVD of the matrix $\mathbf{X}$) yields a collection of L eigenvalues and eigenvectors.

Step 3: Grouping. A particular combination of a certain number r of these eigenvectors determines an r-dimensional subspace $\mathcal{L}_r$ of R^L, $r < L$. The L-dimensional data $\{X_1, \ldots, X_K\}$ is projected onto the subspace $\mathcal{L}_r$.

Step 4: Diagonal averaging. The averaging over the diagonals of the projected data yields a Hankel matrix $\widetilde{\mathbf{X}}$. The time series $(\tilde{x}_1, \ldots, \tilde{x}_N)$, which is in the one-to-one correspondence with the matrix $\widetilde{\mathbf{X}}$, provides an approximation to either the whole series $\mathbb{X}_N$ or a particular component of $\mathbb{X}_N$.

© The Author(s), under exclusive license to Springer-Verlag GmbH, DE,
part of Springer Nature 2020
N. Golyandina and A. Zhigljavsky, *Singular Spectrum Analysis for Time Series*, SpringerBriefs in Statistics,
https://doi.org/10.1007/978-3-662-62436-4_1

Steps 1 and 2 comprise the first stage of Basic SSA called *Decomposition stage*; steps 3 and 4 form the second stage of Basic SSA called *Reconstruction stage*.

Basic SSA and models of time series (*Sect.* 2.3.1). As a non-parametric and model-free method, Basic SSA can be applied to any series. However, for interpreting results of analysis and making decisions about the choice of parameters some models may be useful. The main assumption behind Basic SSA is the assumption that the time series can be represented as a sum of different components such as trend (which we define as any slowly varying time series), modulated periodicities and noise. All interpretable components can be often approximated by time series of small rank and hence can be described via certain linear recurrence relations.

LRR (*Sects.* 2.3.1, 3.2). Consider a time series $\mathbb{X} = (x_1, \ldots, x_N)$ with $N \leq \infty$ and let r be a positive integer. We say that $\mathbb{X}$ is governed by a *linear recurrence relation* (LRR) if $x_n = \sum_{j=1}^{t} a_j x_{n-j}$, for all $n = t + 1, \ldots, N$, where $a_1, \ldots, a_t$ are some real numbers (coefficients of the LRR). If $a_t \neq 0$ then the LRR has order t. LRRs play very important role in the model-based SSA.

Capabilities of Basic SSA (*Sect.* 2.2 *and also Sects.* 3.1, 3.7, 3.8 *and* 3.9). The list of major tasks, which Basic SSA can be used for, includes smoothing, noise reduction, extraction of trends of different resolution, extraction of periodicities in the form of modulated harmonics, estimation of volatility, etc. These tasks are considered in Sect. 2.2. SSA filtering is examined in Sect. 3.9. Forecasting and filling in missing values, the more advanced model-based SSA capabilities, are discussed in Sects. 3.1 and 3.7, respectively. Add-ons to Basic SSA permitting estimation of signal parameters are considered in Sect. 3.8. All major capabilities of Basic SSA are illustrated on real-life time series.

Choice of parameters in Basic SSA (*Sect.* 2.4). There are two parameters to choose in Basic SSA: the window length L and the group of r indices which determine the subspace $\mathcal{L}_r$. A rational or even optimal choice of these parameters should depend on the task we are using SSA for. The majority of procedures require interactive (including visual) identification of components. An automatic choice of parameters of Basic SSA is also a possibility, see Sect. 2.4.5. However, statistical procedures for making this choice are model-based. Success in using the corresponding versions of SSA depends on the adequacy of the assumed models and especially on achieving good separability of the time series components.

Use of prior information (*Sect.* 2.5.2). Basic SSA can be modified and extended in different ways to incorporate prior information. In particular, as a frequently used modification of Basic SSA, in Sect. 2.5.2.2 we consider a common application of Toeplitz SSA for the analysis of stationary time series. (In the literature on SSA, Basic SSA is sometimes called BK SSA and what we call 'Toeplitz SSA' is sometimes called VG SSA; here BK and VG stand for Broomhead & King [10, 11] and Vautard & Ghil [79], respectively.) Under the assumption that the time series $\mathbb{X}_N$ is stationary, the matrix $\mathbf{X}\mathbf{X}^T$ of Basic SSA can be replaced with the so-called lag-covariance matrix $\mathbf{C}$ whose elements are $c_{ij} = \frac{1}{N-k} \sum_{t=1}^{N-k} x_t x_{t+k}$ with $i, j = 1, \ldots, L$ and $k = |i - j|$. Unsurprisingly, if the original time series is stationary then Toeplitz SSA slightly outperforms Basic SSA. However, if the time series is not stationary then Toeplitz SSA may yield results which are simply wrong.

Improving separability (*Sect.* 2.5.3). Achieving separability of time series components, which allows one to analyse them separately, is crucial in the SSA analysis. To improve separability, a variety of SSA modifications have been developed and subsequently implemented in the Rssa package introduced in Sect. 1.4. In Sect. 2.5.3.2, we describe the approach, which uses rotations for improving separability; this approach can be viewed as a combination of SSA and Independent Component Analysis (ICA).

Multivariate and multidimensional extensions of SSA (*Sect.* 2.6). Versions of SSA used for simultaneous analysis and forecasting of several times series are referred to as MSSA (Multivariate or Multichannel SSA) and considered in Sect. 2.6.1. MSSA is a direct extension of the standard 1D-SSA (one-dimensional SSA). 2D-SSA for image processing is considered in Sect. 2.6.2. 2D-SSA is also an extension of 1D-SSA, but it is used for analyzing images. The only difference between the 1D-SSA and 2D-SSA algorithms is in the construction of the trajectory matrix: the moving window in 2D-SSA is a rectangle and the window length is a product of two numbers. This implies, in particular, that the size of the trajectory matrix could be very large and a clever implementation of the SVD becomes essential. The most general version of SSA is called Shaped SSA; it is briefly considered in Sect. 2.6.3.

SSA forecasting (*Sects.* 3.1–3.6). Time series forecasting is an area of huge practical importance and Basic SSA can be very effective for forecasting. The main idea of SSA forecasting is as follows.

Assume $\mathbb{X}_N = \mathbb{X}_N^{(1)} + \mathbb{X}_N^{(2)}$ and we are interested in forecasting of $\mathbb{X}_N^{(1)}$. If $\mathbb{X}_N^{(1)}$ is a time series of finite rank, then it generates some subspace $\mathcal{L}_r \subset \mathsf{R}^L$. This subspace reflects the structure of $\mathbb{X}_N^{(1)}$ and can be taken as a base for forecasting. Under the conditions of separability between $\mathbb{X}_N^{(1)}$ and $\mathbb{X}_N^{(2)}$ (these conditions are discussed throughout the volume; see, for example, Sects. 2.3.3, 2.4, 3.3.1 and 3.5), Basic SSA is able to accurately approximate $\mathcal{L}_r$ and hence it yields an LRR, which can be directly used as a forecasting formula. This method of forecasting is called recurrent forecasting and is considered in Sect. 3.3; it is also revisited at several other sections. Alternatively, we may use the so-called 'vector forecasting'; the main idea of this algorithm is in the consecutive construction of the vectors $X_i = (x_i, \ldots, x_{i+L-1})^{\mathrm{T}}$, for $i = K+1, K+2, \ldots$ so that they lie as close as possible to the subspace $\mathcal{L}_r$.

Short-term forecasting makes very little use of the model while responsible forecasting for long horizons is only possible when an LRR is built by SSA and the adequacy of this LRR is testified. As demonstrated in Sect. 3.3, quality of long-term SSA forecasting is very much determined by the properties of the characteristic polynomials associated with these LRRs. Specifically, the accuracy of SSA forecasting formulas depends on the location of the roots of these polynomials. In Sect. 3.2 we provide an overview of the relations between LRRs, the characteristic polynomials and their roots and discuss properties of the so-called min-norm LRRs, which are used for estimating parameters of the signal (see Sect. 3.8), in addition to forecasting.

In forecasting methodology, construction of confidence intervals for forecasts is often an essential part of the procedure. Construction of such intervals for SSA forecasts is discussed in Sect. 3.4. Despite SSA itself is a model-free technique, for building confidence intervals we need to make certain assumptions such as the assumption that the residual series is a realization of a stochastic white noise process.

In Sect. 3.5 we give recommendations on the choice of forecasting parameters and in Sect. 3.6 we discuss results of a case study. We argue that stability of forecasts is the major aim we must try to achieve in the process of building forecasts. Forecast stability is highly related to the forecast precision and forecast reliability.

SSA for missing value imputation (*Sect.* 3.7). Forecasting can be considered as a special case of missing value imputation if we assume that the missing values are located at the end of the time series. We show how to extend some SSA forecasting procedures (as well as methods of their analysis) to this more general case.

Parameter estimation in SSA and signal processing (*Sect.* 3.8). Although there are many similarities between SSA and the subspace-based methods of signal processing, there is also a fundamental difference between these techniques. This difference is caused by the fact that the model of the form 'signal plus noise' is mandatory in signal processing; consequently, the main aim of the signal processing methods is the estimation of parameters of the model (which is usually the sum of (un)damped sinusoids). The aims of SSA analysis are different (for instance, splitting the time series into components or simply forecasting) and the parameters of the approximating time series are of secondary importance. This fundamental difference between the two approaches leads, for example, to different recommendations for the choice of the window length L: a typical recommendation in Basic SSA is to choose L reasonably large whereas in respective signal processing methods the value of L is typically relatively small. This also concerns Hankel SLRA (structured low-rank approximation) discussed in Sect. 3.8.4.

SSA and filters (*Sect.* 3.9). One of interpretations of Basic SSA is related to its association with linear filters (which are constructed in a non-linear way!). Connections between SSA and filtering are considered in detail in Sects. 3.9.2, 3.9.3 and 3.9.4. In Sect. 3.9.5, we examine the so-called Causal SSA, which originates from considering SSA as the last-point filter of Sect. 3.9.4 and can be considered as an alternative to SSA forecasting. In Causal SSA, we assume that the points in the time series $\mathbb{X}_\infty = (x_1, x_2, \ldots)$ arrive sequentially, one at a time. Starting at some $M_0 > 0$, we apply Basic SSA with fixed window length and the same grouping rule to the time series $\mathbb{X}_M = (x_1, \ldots, x_M)$ for all $M \geq M_0$. We then monitor how SSA reconstruction of previously obtained points of the time series change as we increase M (this is called redrawing). The series consisting of the last points of the reconstructions is the result of Causal SSA. The delay of the Causal SSA series reflects the quality of forecasts based on the last points of the reconstructions. Additionally to the redrawings of the recent points of the reconstructions, this delay can serve as an important indicator of the proper choice of the window length, proper grouping and, which is very important, of predictability of the time series.

Model-based applications for multivariate and multidimensional versions of SSA (*Sect.* 3.10). Many considerations of Chap. 3, formulated for one-dimensional time series, can be extended for the multivariate/multidimensional versions of SSA introduced in Sect. 2.6. This is briefly discussed in Sect. 3.10; much more material on this topic can be found in [31].

1.2 SSA-Related Topics Outside the Scope of This Book

The book covers the majority of topics related to the methodology of SSA. There are, however, a few important SSA-related topics which are covered insufficiently or not covered at all. Let us briefly consider them.

Theory of SSA. For the basic theory of SSA we refer to the monograph [29]. Since the publication of that book, several influential papers on theoretical aspects of SSA have been published. The main theoretical paper on perturbations in SSA and subspace-based methods of signal processing is [60], see a related discussion at the end of Sect. 2.3.3. Another important theoretical paper is [78], where the concept of SSA separability is developed further (relative to [29]) and studied through characteristics of LRRs. Elements of the theory of SSA are also discussed in [25]. Other theoretical aspects of SSA deal with parameter estimation and Hankel SLRA, see Sect. 3.8.

SSA for change-point detection and subspace tracking. Assume that the observations $x_1, x_2, \ldots$ of the time series arrive sequentially in time and we apply Basic SSA to the observations at hand. Then we can monitor the distances from the sequence of the trajectory matrices to the r-dimensional subspaces we construct and also the distances between these r-dimensional subspaces. Significant changes in any of these distances may indicate a change in the mechanism generating the time series. Note that this change in the mechanism does not have to affect the structure of the whole time series but of a few of its components only. For some references, we refer to [59, 62] and [29, Chap. 3].

Numerical comparison of SSA with other methods of time series analysis and forecasting. Numerical comparison of SSA with ARIMA and other classical methods of time series analysis can be found in several papers of the volume [87] and in many papers devoted to applications of SSA, see for example [4, 19, 36, 65, 75].

SSA and machine learning. In many applications, SSA is incorporated as a part of machine learning algorithms; for example, SSA can be used for feature extraction or, visa versa, machine learning techniques can be used for the choice of SSA parameters. In [31, Sect. 1.7.3] one can find a brief review of papers, where SSA is combined with machine learning techniques such as SVM, neural nets, random forest, and others.

Application areas. SSA has proved to be very useful and has become a standard tool in the analysis of climatic, meteorological and geophysical time series; see, for example, [22, 73, 79] (climatology), [44, 77, 82] (meteorology), [17] (marine science), [48] (geophysics), [67] (seismology); for more references, see [1, 3, 18, 21, 29, 31, 54, 80, 81]. More recent areas of application of SSA include biomedicine [68, 83], soil science [52], study of Earths magnetic field [76], hyperspectral imaging [85, 86], prediction of Earth orientation parameters [63], blind source separation

[56], chemistry [40], traffic surveillance [15], cryptography [23], robotics [58], GPS navigation [35], tourism [19, 39, 72], social nets [20], among many other fields. A special case is econometrics, where SSA was basically unknown 15 years ago but it has made a tremendous advancement there and is becoming more and more popular; see, for example, [14, 36–38, 65].

1.3   SSA and Other Techniques

This section is mostly based on a recent survey [27].

1.3.1   Origins of SSA and Similar Techniques

Karhunen–Loève transform. To a large extent, the structure of a random process ξ_t ($t \in [0, T]$) is contained in its autocovariance function $K(s, t) = \mathsf{E}(\xi_s - \mathsf{E}\xi_s)(\xi_t - \mathsf{E}\xi_t)$. Assuming that ξ_t has zero mean ($\mathsf{E}\xi_t = 0$, $\forall t$), the Karhunen–Loève transform (KLT) is the decomposition of ξ_t into an infinite sum $\xi_t = \sum_k u_k(t)\varepsilon_k$, where $u_k(t)$ are eigenfunctions of the linear integral operator with kernel $K(s, t)$ and $\varepsilon_1, \varepsilon_2, \ldots$ are uncorrelated random variables (white noises).

For discrete time, the KLT of $\mathbb{X}_N = (x_1, \ldots, x_N)$ coincides with PCA (Principal Component Analysis) of the random vector $\xi = (\xi_1, \ldots, \xi_N)^\mathsf{T}$. Therefore, the KLT is the decomposition of ξ into a finite sum of white noises ε_k, $k = 1, \ldots, N$, with coefficients obtained from the eigenvectors U_k of the covariance matrix of ξ: $\xi = \sum_{k=1}^N U_k \varepsilon_k$. If $\mathbb{X}_N = (x_1, \ldots, x_N)$ is assumed to be a realization of $(\xi_1, \ldots, \xi_N)$, then the empirical KLT is constructed on the basis of eigenvectors of the sample covariance matrix of ξ. If the time series $\mathbb{X}_N$ is centered or, alternatively, centering is applied to the rows of the trajectory matrix $\mathbf{X}$, then $\mathbf{X}\mathbf{X}^\mathsf{T}$ can be considered as an estimate of the covariance matrix of ξ. The question now is how the sample version of ε_k ($k = 1, \ldots, N$) is constructed. In PCA, we have many samples from $\xi = (\xi_1, \ldots, \xi_N)^\mathsf{T}$ and the sample version of ε_k is the k-th principal component of the multivariate data. For time series $\mathbb{X}_N$, we have only one realization of length N. By constructing the trajectory matrix $\mathbf{X}$, we create $K = N - L + 1$ L-dimensional samples so that the orthogonal vectors $\mathbf{X}^\mathsf{T} U_m \in \mathsf{R}^K$ are considered as sample versions of ε_m, $m = 1, \ldots, \min(L, K)$. Therefore, the empirical KLT (see [6] for a full description of the empirical KLT) formally corresponds to the Decomposition stage of SSA with centering; this stage is considered in detail in Sects. 2.1.1.1 and 2.5.2.1.

Specialists in the theory of random processes, who have briefly familiarized themselves with SSA, sometimes think that SSA is simply the empirical KLT. This is not the case for the following three reasons: (i) the Decomposition stage is only a part of the SSA algorithm and the Reconstruction stage is not a part of the empirical KLT, (ii) centering is not mandatory in SSA, and (iii) the view on SSA from the position of the theory of random processes leaves a mark on the methodology of applications of SSA (see, e.g., [47]).

Theory of stationary time series. For specialists in stationary time series analysis, it is natural to assume that the data $\mathbb{X}_N = (x_1, \ldots, x_N)$ is a realization of a stationary discrete-time process and Toeplitz SSA of Sect. 2.5.2.2 is the version of SSA such specialists would choose. This approach to SSA significantly limits the range of SSA applications to the analysis of realizations of random stationary processes only, whereas the range of problems solved by SSA is much wider. Moreover, when at least one of the two assumptions (of randomness and of stationarity) fails, Toeplitz SSA may lead to wrong interpretations and conclusions. Since this version of SSA has been proposed for the analysis of climatic data as the default option [79], it is still considered as the main version of SSA in climatology and related areas. This seems to be in conflict with that Basic SSA is a much easier tool for separating various effects of global warming (assuming it is real) from seasonality and other effects.

Takens embedding theorem in dynamical systems. Basic SSA has been properly formulated and received its name in papers [10, 22], which were dealing with the Takens embedding theorem in dynamical systems. As a result, some terminology in SSA is taken from the theory of dynamical systems. The dynamical system way of approaching the problem has left the term 'embedding' for the first step of the SSA algorithm as well as the term 'trajectory matrix'. From the modern prospective, the connection between SSA and the Takens theorem is mostly historical (and hence we omit the description of the Takens' 'embedology' approach, see [70] for details). A discussion on reliability of the Broomhead&King method of estimating the embedding dimension can be found in [55].

Structured low-rank approximation (SLRA). A particular case of the Basic SSA algorithm, when extracting the signal is of interest, coincides with one step of the iterative algorithm suggested in [12] and often referred to as 'Cadzow iterations'. The aim of Cadzow iterations is to construct the approximation of the trajectory matrix by a Hankel low-rank matrix, which is a problem of Hankel SLRA, see Sect. 3.8.4. The matrix called 'trajectory' in SSA, was called 'enhanced matrix' in the Cadzow's paper and some subsequent papers: for example, [42] constructs the enhanced matrix in the 2D case for 2D frequency estimation. Basic SSA and Hankel SLRA are designed for solving different problems, see discussion in [27].

1.3.2 Is SSA a Linear Method?

Some authors (see, for example, the first sentences in [41, 61]) do not consider SSA as a serious technique on the base of their belief that SSA is a linear method. This belief is based on the observation that the class of time series producing rank-deficient trajectory matrices consists of the time series governed by homogeneous LRRs in the form $x_n = \sum_{j=1}^{r} a_j x_{n-j}$. LRRs are closely related to linear differential equations (LDEs) $\sum_k b_k s^{(k)}(t) = 0$, where $s^{(k)}(t)$ is the k-th derivative of $s(t)$:

in particular, LRRs are generated by the finite-difference method applied for solving linear differential equations. In the theory of dynamical systems, the methods, which are related to linear differential equations, are called linear. For example, let $s(t) + bs'(t) = 0$, $t \in [0, \infty)$, be a LDE of order 1; consider $j = 1, 2, 3, \ldots$ as the discretization of t and the finite-difference approximation $s'(j) \approx s_{j+1} - s_j$; then we obtain $s_j + b(s_{j+1} - s_j) = 0$ or $s_{j+1} = (1 - 1/b)s_j$, which is an LRR of order 1. Note that the class of solutions of LDEs is rather wide and it contains arbitrary finite sums of products of polynomials, exponentials and sine waves.

Let now us give some arguments against the claim that SSA is a linear method.

(1) SSA does not use an explicit parametric form of the time series. For example, to predict the value of an exponential series $s_n = A e^{\alpha n}$, one approach is to estimate A and α and then perform forecasting by an explicit formula. However, the SSA forecasting algorithms use a more flexible approach consisting of estimating coefficients of a governing LRR and using them for forecasting. This way, most versions of SSA avoid the use of parametric forms of time series and thus make the method more robust with respect to deviations from the model.

(2) SSA with small window lengths. To deal with signals governed by LRRs (e.g., to extract signals from noisy time series), a large window length (optimally, in the range from $N/3$ to $N/2$, where N is the time series length) is recommended, see Sect. 2.4.3; at the same time, smaller windows allow taking into consideration local finite-rank approximations. For example, a modulated sinusoid with a slowly varying amplitude is well approximated by a sinusoid with the same frequency on time intervals of size equal to several periods. In this example, SSA with a small window length equal to several periods achieves reasonable results. It can be argued that in order to be extracted by SSA, the time series component of interest should be locally governed by the same LRR at each local segment. Therefore, SSA with small window lengths allows us to deal with nonlinearity of the model in some common cases, such as modulated harmonics and trends. Note, however, that while a slowly changing amplitude of the sinusoidal components is admissible, a changing frequency is not appropriate for SSA applied to the whole time series (except in the case of 'phase noise', when the frequency is varying around one main value). For extraction of oscillations with changing frequency, methods like EMD + HHT [43] can be used but they work well only in cases when the signal/noise ratio is high.

(3) Local SSA. The standard approach for analyzing signals with a changing structure is to consider moving subseries (segments) of the original time series. In the framework of SSA, this procedure is called subspace tracking. The main attention in many papers on subspace tracking is on the construction of fast algorithms. One of the main applications of subspace tracking is structure monitoring and change-point detection; the topic mentioned in Sect. 1.2.

The main problem, when SSA is applied to moving subseries, is defining the rule for combining different local decompositions into one global decomposition. In [53], central parts of the local signal reconstructions are used and then stacked. The suggestion of using the central parts only is very sensible as the reconstruction of the end-points is less accurate. However, the question is why not to use only one central point (by analogy with the LOESS method [16]); this can be easily done if the computational cost is acceptable. As a result of application of local SSA, we obtain a signal estimation or simply a smoothed time series. The problem is how to forecast the extracted signal, since its local estimates may have different structures on different time intervals. Indeed, by using local versions of SSA, we do not obtain a common nonlinear model but instead we have a set of local linear models. Many nonparametric local methods including LOESS have this difficulty.

(4) SSA as a linear filter. One of the reasons for thinking about SSA as a linear method is its connection with linear filters, see Sect. 3.9. Note, however, that the coefficients in the SSA linear filters are computed in a nonlinear manner.

1.3.3 SSA and Autoregression

The basic model of signals in SSA and the model of an autoregressive process (AR) look similar but in fact they are totally different. In SSA, the signal model is an LRR $s_n = \sum_{j=1}^{r} a_j s_{n-j}$ and the observed time series is a noisy signal: $x_n = s_n + \epsilon_n$, where ϵ_n is a noise such as a non-regular oscillation or a realization of a stationary random process. In the AR model, we have noise innovations at each step: $x_n = \sum_{j=1}^{r} a_j x_{n-j} + \epsilon_n$. Under some conditions on the initial values $x_1, \ldots, x_r$ and the coefficients $a_1, \ldots, a_r$, such innovations yield that $\{x_n\}_{n=1}^{N}$ is a stationary discrete-time stochastic process, while the time series $\{s_n\}_{n=1}^{N}$ with terms $s_n = \sum_{j=1}^{r} a_j s_{n-j}$ is deterministic and not necessarily stationary. The coefficients $\{a_j\}_{j=1}^{r}$ in the SSA model can be arbitrary.

There are, however, some connections between SSA and AR analysis. In both AR and SSA analysis, the characteristic polynomials are constructed on the basis of the coefficients $a_1, \ldots, a_r$. In SSA, we are concerned about characteristic roots of the governing LRR, that is, about the roots of the characteristic polynomial $\mu^r - a_1 \mu^{r-1} - \ldots - a_r$ (see Sect. 3.2). The roots, which have moduli larger/smaller than 1, correspond to a growing/damped time series components, whereas roots with unit moduli correspond to stationary deterministic components like undamped sine waves. The AR characteristic polynomial is reciprocal to that of SSA (the coefficients are taken in the reverse order); therefore, its roots are inverse to the roots of the characteristic polynomial in SSA with the same sets of coefficients $a_1, \ldots, a_r$. The stationarity in the AR model corresponds to the case when all the roots of the AR characteristic polynomial have moduli larger than 1. The case of roots with unit moduli imply non-stationarity (an example is Brownian motion $x_n = x_{n-1} + \epsilon_n$ with characteristic polynomial $\mu - 1$; the same characteristic polynomial corresponds to a constant signal in SSA).

There is a common form for forecasts of signals governed by LRRs and for conditional mean forecasts for the AR processes. Both are performed by the LRR with estimated coefficients. The difference is in the approach for estimating these coefficients. Note that the forecasting values in the stationary AR models are always converging to zero (or the mean value).

For the signal model consisting of a trend and seasonality, Seasonal ARIMA can be competitive with SSA. There are many real-world examples where SSA provides better accuracy than (Seasonal) ARIMA and vice versa; see references in Sect. 1.2. Note that the Seasonal ARIMA model requires the knowledge of the period of the periodic time series component whereas SSA does not. Moreover, a time series of length 2–3 periods can be successfully analysed with SSA but analyzing such short time series with Seasonal ARIMA is usually hopeless. One of the advantages of ARIMA is the ability to automatically select the order of the ARIMA model using information criteria like the Akaike information criterion (AIC) or the Bayesian information criterion (BIC). In SSA, the number r of signal components can also be chosen on the basis of the AIC/BIC approach; however, as discussed in Sect. 2.4.5, this approach has limitations.

There is one more connection between SSA and AR. For autoregressive processes, SSA can be applied for estimation of the autocovariance function $C(l)$ for finding the autoregressive coefficients, since the series $C(l)$ is governed by the LRR with the same coefficients: $C(l) = \sum_{j=1}^{r} a_j C(l - j)$ (the Yule–Walker equations).

In the context of SSA used for signal extraction, AR is mostly considered as a noise model (see, e.g., a discussion below on Monte Carlo SSA). Let us also point out that there exists at least one hybrid of SSA and AR [46]: the AR model is used for forecasting leading elementary reconstructed components obtained by the Toeplitz (VG) version of SSA. In [46], this method was used for forecasting several sets of climate data, which were assumed to be realizations of stationary random processes.

1.3.4   SSA and Linear Regression for Trend Estimation

As discussed above, SSA can extract trends, which are slowly-varying time series components; such components can often be accurately approximated by finite-rank time series. Ideally (but not necessarily), trends should be time series of small rank r. Linear series with terms $s_n = an + b$, $a \neq 0$, $n = 1, \ldots, N$, belong to this class of time series, and have rank 2. However, linear functions are not natural for SSA (while exponential series are very natural). The reason is that the characteristic root for a linear series is 1 and has multiplicity 2. Indeed, the minimal LRR is $s_{n+2} = 2s_{n+1} - s_n$ and, therefore, the characteristic polynomial is $\mu^2 - 2\mu^2 + 1$. The presence of roots of multiplicity larger than 1 produces instability. Nearly any distortion of coefficients of the LRR $s_{n+2} = 2s_{n+1} - s_n$ transforms the multiple unit root into two different roots; that is, a linear series is transformed either into a sum of two exponentials with small exponential rates or into a sinusoid with large period, see Example 3.3.

There is a modification of SSA, which was called in [29, Sect. 1.7.1] 'SSA with double centering'. In [29], a correspondence between SSA with double centering and extraction of linear trends is demonstrated. As shown in [28], SSA with double centering is a particular case of SSA with projection, where projections of rows and columns of the trajectory matrix to given subspaces are performed and then subtracted from the trajectory matrix. After that, the SVD expansion of the residual matrix is made. SSA with projection is positioned as SSA with the use of prior information, see Sect. 2.5.2. Until now, SSA with projection has been analyzed only for the case where extraction of polynomial trends is the objective of a study.

The most common approach for estimating linear trends is the linear regression method, where the least-squares (LS) solution is used. Let us summarize the results of comparison of SSA with double centering and linear regression. Certainly, if the time series consists of a linear trend and white noise (that is, $x_n = an + b + \epsilon_n$), then the LS method provides the best estimator of the trend. However, if the residual includes, e.g., a periodic component, this is not true. There is plenty of numerical evidence reported in [28] that for the case $x_n = an + b + \sin(2\pi\omega n + \phi) + \epsilon_n$, where ϵ_n is white noise, the LS method applied to the trend, which was obtained by SSA with double centering, is superior to the conventional LS estimate applied to the original time series.

1.3.5 SSA and DFT, EMD, DWT

DFT. Discrete Fourier transform (DFT) differs from SSA by the use of a fixed basis consisting of sines-cosines with frequencies from an equidistant grid; in contrast, SSA constructs an adaptive basis. In [8], a relation between SSA and DFT is discussed. Note that the SSA model suits DFT in the circular version of SSA (see the terminology in [71]), where the data is given on a circle in the 1D case and on a torus in the 2D case.

From the viewpoint of frequency estimation (see Sect. 3.8), SSA and the related subspace-based methods allow estimation of frequencies with a better resolution than $1/N$, where N is the time series length (see, e.g., [69, 74] for comparisons of DFT, ESPRIT and MUSIC). The MUSIC method allows construction of pseudo-spectrums similar to the periodograms but with no limitation on the frequency set. By comparing time series models that are suitable for DFT and SSA, we can conclude that a sum of pure sinusoids corresponds to DFT while a sum of exponentially modulated sinusoids corresponds to SSA.

Another application of DFT is the estimation of the spectral density by means of smoothing the periodograms. SSA can be used for estimating the spectral density, see, e.g., [29, Sect. 6.4.3], where the results from [33] in terms of SSA are discussed. If the spectral density is monotonic, different eigenvectors generally correspond to different frequencies; otherwise, the eigenvectors are mixed, i.e. they may include different frequencies with comparable contributions. This explains why most eigenvectors produced by white noise (which has a constant spectral density) are irregularly oscillating. On the other hand, the eigenvectors generated by red noise (the autoregressive process of order one with a positive coefficient) correspond to different frequencies.

EMD. Empirical mode decomposition (EMD) [43] is frequently compared with SSA, since both are model-free techniques. It seems that EMD is a method without explicit approximation properties, whereas SSA has both separability and approximation properties. The first components extracted by EMD are highly oscillating and the last component is considered as a trend. It is the opposite for SSA, since the signal and the trend typically correspond to the leading components of the SSA decomposition; this is an advantage of SSA as a decomposition method. The advantage of EMD is its ability to extract periodic components with complex amplitudes and frequency modulations. There are papers (e.g., [57]), where a combination of SSA and EMD is used for solving a real-world problem; in [85], an superiority of SSA over EMD is demonstrated.

DWT. Discrete wavelet transform is the decomposition based on a fixed space-time basis. This yields both advantages and disadvantages in comparison with SSA. See discussion in [84].

1.3.6   SSA and Signal Detection; Monte Carlo SSA

The problem of signal detection is very important in practice. If noise is strong, it is hardly surprising that many methods tend to discover spurious signals, since noise (being considered as a stationary process) contains all frequencies. The mean contribution of each frequency is determined by the spectral density of noise. In particular, this implies that if the spectral density is larger for low frequencies, the probability of spurious trends or spurious sine waves with low frequencies increases. This is exactly the case of the so-called red noise; that is, an autoregressive process of order 1, AR(1), with a positive coefficient smaller than 1.

Since we observe one realization of a time series, the contribution of each frequency from the grid $\{k/N, \ k = 0, \ldots, \lfloor N/2 \rfloor\}$ (we consider the periodogram values, which correspond to these frequencies, as their contributions) is random with variance, which does not tend to zero as N tends to infinity. Moreover, it has an exponential distribution implying that large values are likely.

The question of presence of a signal in noise can be reduced to the construction of a criterion for testing the null hypothesis that the time series is pure noise; the criterion should be powerful against the alternative that a signal is present. There are many such criteria for different models of noise. Most of them deal with the white noise model.

In the framework of SSA, red noise is the other focus of attention. One of the reasons for this is that historically, SSA was primarily popular in climatology, where red noise is a conventional model of a climatic time series. Moreover, the properties of red noise suit SSA, since red noise has a monotonic spectral function; see a brief discussion in Sect. 1.3.5.

The method for detection of a signal in red noise is traditionally called 'Monte Carlo SSA' [2, 3, 32, 34, 45, 64], for the very simple reason that it uses simulations. The name of the method (Monte Carlo SSA) does not adequately reflect its purpose, which is a hypothesis testing; nevertheless, that is the name by which this method is known. The approach used in Monte Carlo SSA is straightforward. First, a characteristic of data that reflects a disparity between the null and alternative hypotheses should be chosen; then, using simulations, surrogate data is generated according to the null-hypothesis in order to construct the distribution of the chosen characteristic. Finally, one checks if the chosen characteristic of the real-world data, which can be called test statistic, lies outside $\alpha/2$- and $(1 - \alpha/2)$-quantiles of the constructed distribution. If this is the case, then the null hypothesis is rejected at the significance level α. In the case of Monte Carlo SSA, this characteristic is the squared norm of the projection of the trajectory matrix on a chosen vector which refers to a given frequency. The relation with SSA is in the choice of the projection vector as one of the eigenvectors of the trajectory matrix; then the test statistic is equal to the corresponding eigenvalue.

Certainly, there are additional problems, which should be solved in the course of application of the described scheme. For example, parameters of the AR(1) process satisfying the null-hypothesis are unknown and should be estimated. For signal detection, it is not enough to choose only one characteristic (in Monte Carlo SSA, one projection vector corresponds to one frequency), since a hypothetic signal contains frequencies unknown in advance. Thus, the problem of multiple testing arises. We refer to [26] for the description of a more strict statistical approach for constructing the Monte Carlo SSA test, where the problem of multiple testing is solved and an approach for controlling the type I error and estimating the test power is developed.

1.4 Computer Implementation of SSA

There are many implementations of SSA. They can be classified as follows. First, the implementations differ by the potential areas of application: for example, general purpose SSA, see e.g. [29], and SSA oriented mainly for climatic applications, see e.g. [24]. Second, the software can be either free-access or not free-access. One of the main drawbacks of free-access packages is that they generally have no support and that their implementation consists of direct and usually non-efficient use of the main formulas. Third, the software can be interactive (for different systems, Window, Unix or Mac) and non-interactive. Interactive implementations of SSA provide executable programs in some programming language such as special mathematical languages like R and Matlab or high-level programming languages like C++, VBA and others. We draw special attention to the following four supported software packages:

1. http://gistatgroup.com:
   'Caterpillar'-SSA software (Windows) following the methodology from [29];
2. http://www.atmos.ucla.edu/tcd/ssa/:
   SSA-MTM Toolkit for spectral analysis [24] (Unix) and its commercial extension kSpectra Toolkit (Mac);
3. http://cran.r-project.org/web/packages/Rssa/:
   R-package Rssa [29, 31, 49, 51], a very fast implementation of the main SSA procedures for any platform.

4. The commercial statistical software, SAS, includes SSA to its econometric extension SAS/ETS®Software.

The fastest implementation of SSA can be found in the R-package Rssa. Let us briefly describe the idea of its implementation [30, 31, 49]. The most time-consuming step of SSA is the Singular Value Decomposition (SVD). The SVD in SSA has two specific features. First, SSA as a rule uses only a few leading components. Therefore, we need to use the so-called Partial SVD to compute only a given number of leading eigentriples. Second, the trajectory matrix used for decomposition is Hankel. This feature of the trajectory matrix can be effectively used to speed up the matrix-vector multiplications. The fastest acceleration is reached for the case $L \sim N/2$, which is a commonly used window length, and for large N. However, even for moderate N the advantage is often very visible. The acceleration in the Rssa package is achieved by the following means.

- The embedding step is combined with the SVD step; this decreases the storage requirement as we do not need to store the trajectory matrix.
- The Rssa package includes the Partial SVD that generally provides the computational complexity $O(rN^2)$ for calculating r eigentriples rather than $O(N^3)$ needed for the full SVD. The Lanczos-based version of the Partial SVD implemented in the R-package svd [50] performs computations with the help of an external function of multiplication of the decomposed matrix by a vector. The use of the Fast Fourier Transform (FFT) for the multiplication of a Hankel matrix by a vector decreases the computational complexity of the SVD step to $O(rN \log N)$.
- Similarly, FFT is used at the Reconstruction stage, in particular, for fast hankelization; this reduces its complexity for reconstruction of r components from $O(rN^2)$ to $O(rN \log N)$.

Summarizing, very fast effective algorithms are implemented in the Rssa package with the computational cost (in flops) of the SSA algorithm dropped from $O(N^3)$ down to $O(rN \log N + r^2 N)$ and the memory consumption reduced from $O(N^2)$ to $O(N)$; here r is the number of calculated leading eigentriples and the window length L is assumed to be proportional to N.

A very similar approach is implemented in Rssa for decreasing the computational cost and the memory requirements in the multidimensional versions of SSA, [30, Sect. 6.2]; here N stands for the number of elements in the original object.

There are different approaches for fast SVD updating (e.g., a so-called incremental SVD, or online PCA) [5, 9, 13] with linear in N complexity for updates. However, algorithms of incremental SVD with linear computational cost generally do not provide exact SVD for the case of calculating only r leading SVD components, if r is smaller than the rank of the decomposed matrix. Therefore fast incremental SVD algorithms cannot be incorporated in SSA without certain loss of precision.

Example of calculations in Rssa. Let us demonstrate how fast are the computations in Rssa. For the time series length $N = 10^6$ and the window length $L = N/2$, the reconstruction of a signal of rank 3 based on three leading components is executed in a few seconds (processor Intel(R) Core(TM) i7-8550U CPU

@ 1.80 GHz 1.99 GHz, 64-bit system); see the processing time in seconds in the output of system.time:

```
> library("Rssa")
> N <- 1000000
> set.seed(1)
> signal <- 1 + sin((1:N) * 2 * pi / 10)
> ts <- signal + 10 * rnorm(N)
> system.time(s <- ssa(ts, L = N %/% 2, neig = 3))
        user        system       elapsed
        1.22          0.20          1.43
> system.time(rec <- reconstruct(s, groups = list(sig = 1:3)))
        user        system       elapsed
        0.81          0.08          0.89
> max(abs(signal - rec$sig))
\cite{ch1Allen.Robertson1996} 0.05311553
```

Execution of the decomposition was possible since the required memory is $O(N)$ and the algorithm does not need storing the trajectory matrix with $L(N- L+1) \approx 25 \cdot 10^{10}$ elements; storage of this matrix would have required 1.8 terabytes of memory.

1.5 Historical and Bibliographical Remarks

The first publication, which can be considered as one of the origins of SSA (more generally, of the subspace-based methods of signal processing), can be traced back to the eighteenth century [66]. The commencement of SSA is usually associated with publication in 1986 of the papers [10, 11] by Broomhead and King and [22] by Fraedrich; this is despite the fact that the algorithm of SSA was formulated earlier, see e.g. [17]. SSA became widely known in climatology after publication of [79, 80]. The book [21] summarized the basic information about SSA existing to that moment. It is a well-written book but it only provides a very elementary introduction to SSA reflecting the level of SSA-related knowledge at the time of its publication. Moreover, by comparing SSA-related publications of the twentieth century with more recent ones, we have to come to the conclusion that there is a fast progress in understanding of what SSA is and how to use it effectively and reliably.

In parallel, SSA (under the name 'Caterpillar' or 'Gusenitsa', which is due to the association with the moving window) was created in the former Soviet Union independently of the main-stream SSA (the 'iron curtain effect'). The history of the 'Caterpillar' method starts from [7], where the method was adequately formulated. The methodology and theory of the Caterpillar-SSA has been summarized in [18] (in Russian). We believe that a breakthrough in the theory and methodology of SSA has been made in [29], where the authors have significantly developed the theory and the main methodological principles sketched in [18]. After publication of [29], the number of SSA-related publications increased dramatically. The fast increase in the number of new applications of SSA is a consequence of the new fundamental research on SSA and the recent progress in computing and software development; this made it possible to use SSA for very complicated tasks that were simply inconceivable in the twentieth century.

The research on the theory and methodology of SSA performed in the last three decades has resulted in a rather pleasing state of affairs: (i) existence of an active SSA community and (ii) existence of a general methodology of SSA rather than simply a collection of several different SSA algorithms. This methodology unifies different versions of SSA into a very powerful tool of time series analysis and forecasting. Careful examination of the SSA methodology is the purpose of this book.

1.6   Common Symbols and Acronyms

SSA	singular spectrum Analysis
SVD	singular value decomposition
LRR	linear recurrence relation
$\mathbb{X}$ or $\mathbb{X}_N$	time series (or simply 'series')
$\mathbb{X}_N = (x_1, \ldots, x_N)$	time series of length N
$\mathbb{X}_\infty = (x_1, x_2, \ldots)$	infinite time series
N	length of time series
L	window length
$K = N - L + 1$	the number of L-lagged vectors obtained from $\mathbb{X}_N$
$X_i = (x_i, \ldots, x_{i+L-1})^{\mathrm{T}}$	i-th L-lagged vector obtained from $\mathbb{X}_N$
$\mathbf{X} = [X_1 : \ldots : X_K]$	trajectory matrix with columns X_i
$\|\mathbf{X}\|_{\mathrm{F}}$	Frobenius matrix norm
rank $\mathbf{X}$	rank of the matrix $\mathbf{X}$
$\mathfrak{X} = \mathfrak{X}^{(L)}(\mathbb{X}_N)$	L-trajectory space of a time series $\mathbb{X}_N$
$\mathrm{rank}_L(\mathbb{X}_N)$	L-rank of a time series $\mathbb{X}_N$
$\mathfrak{T}$	embedding operator
$\mathbf{\Pi}_{\mathcal{H}}$	hankelization operator
λ_i	i-th eigenvalue of the matrix $\mathbf{X}\mathbf{X}^{\mathrm{T}}$
U_i	i-th eigenvector of the matrix $\mathbf{X}\mathbf{X}^{\mathrm{T}}$
$V_i = \mathbf{X}^{\mathrm{T}} U_i / \sqrt{\lambda_i}$	i-th factor vector of the matrix $\mathbf{X}$
$(\sqrt{\lambda_i}, U_i, V_i)$	i-th eigentriple of the SVD of the matrix $\mathbf{X}$
$\mathbf{I}_M$	identity $M \times M$ matrix
R^L	Euclidean space of dimension L
$\mathcal{L}_r$	r-dimensional linear subspace of R^L
$\mathrm{span}(P_1, \ldots, P_n)$	linear subspace spanned by vectors $P_1, \ldots, P_n$
$\rho^{(w)}$	weighted correlation between two series
Π_x^N	periodogram of a time series $\mathbb{X}_N$

References

1. Alexandrov T, Golyandina N, Spirov A (2008) Singular spectrum analysis of gene expression profiles of Early Drosophila embryo: exponential-in-distance patterns. Res Lett Signal Process 2008:1–5
2. Allen MR, Robertson AW (1996) Distinguishing modulated oscillations from coloured noise in multivariate datasets. Clim Dynam 12(11):775–784
3. Allen MR, Smith LA (1996) Monte Carlo SSA: detecting irregular oscillations in the presence of colored noise. J Clim 9(12):3373–3404
4. Azulay DO, Brain P, Sultana SR (2011) Characterisation of very low frequency oscillations in laser Doppler perfusion signals with a singular spectrum analysis. Microvasc Res 81(3):239–244
5. Baker C, Gallivan K, Van Dooren P (2012) Low-rank incremental methods for computing dominant singular subspaces. Linear Algebra Appl 436(8):2866–2888. Special Issue dedicated to Danny Sorensen's 65th birthday
6. Basilevsky A, Hum DPJ (1979) Karhunen-Loéve analysis of historical time series with an application to plantation births in Jamaica. J Am Stat Assoc 74:284–290
7. Belonin MD, Tatarinov IV, Kalinin OM, Shimanskij VK, Beskrovnaya OV, Granskij VV, Pohitonova TE (1971) Faktornyj analiz v neftyanoj geologii: Obzor [Factor Analysis in Petrolium Geology: Review]. Moskow, VIEMS (in Russian)
8. Bozzo E, Carniel R, Fasino D (2010) Relationship between singular spectrum analysis and Fourier analysis: theory and application to the monitoring of volcanic activity. Comput Math Appl 60(3):812–820
9. Brand M (2006) Fast low-rank modifications of the thin singular value decomposition. Linear Algebra Appl 415(1):20–30. (Special Issue on Large Scale Linear and Nonlinear Eigenvalue Problems)
10. Broomhead D, King G (1986a) Extracting qualitative dynamics from experimental data. Phys D 20:217–236
11. Broomhead D, King G (1986b) On the qualitative analysis of experimental dynamical systems. In: Sarkar S (ed) Nonlinear phenomena and chaos. Adam Hilger, Bristol, pp 113–144
12. Cadzow JA (1988) Signal enhancement: a composite property mapping algorithm. IEEE Trans Acoust 36(1):49–62
13. Cardot H, Degras D (2018) Online principal component analysis in high dimension: which algorithm to choose? Int Stat Rev 86(1):29–50
14. de Carvalho M, Rodrigues PC, Rua A (2012) Tracking the US business cycle with a singular spectrum analysis. Econ Lett 114(1):32–35
15. Chen N, Yang Z, Chen Y, Polunchenko A (2017) Online anomalous vehicle detection at the edge using multidimensional SSA. In: 2017 IEEE conference on computer communications workshops (INFOCOM WKSHPS), IEEE, pp 851–856
16. Cleveland WS (1979) Robust locally weighted regression and smoothing scatterplots. J Amer Stat Ass 74(368):829–836
17. Colebrook JM (1978) Continuous plankton records – zooplankton and environment, northeast Atlantic and North Sea, 1948–1975. Oceanol Acta 1:9–23
18. Danilov D, Zhigljavsky A (eds) (1997) Principal components of time series: the "Caterpillar" method. St. Petersburg Press, St. Petersburg (in Russian)
19. De Klerk J (2015) A comparison of singular spectrum analysis forecasting methods to forecast South African tourism arrivals data. Stud Econ Econ 39(2):21–40
20. Del Vicario M, Vivaldo G, Bessi A, Zollo F, Scala A, Caldarelli G, Quattrociocchi W (2016) Echo chambers: Emotional contagion and group polarization on facebook. Sci Rep 6(37):825
21. Elsner JB, Tsonis AA (1996) Singular spectrum analysis: a new tool in time series analysis. Plenum
22. Fraedrich K (1986) Estimating dimensions of weather and climate attractors. J Atmos Sci 43:419–432

23. Genkin D, Pachmanov L, Pipman I, Tromer E, Yarom Y (2016) ECDSA key extraction from mobile devices via nonintrusive physical side channels. In: Proceedings of the 2016 ACM SIGSAC conference on computer and communications security, pp 1626–1638

24. Ghil M, Allen RM, Dettinger MD, Ide K, Kondrashov D, Mann ME, Robertson A, Saunders A, Tian Y, Varadi F, Yiou P (2002) Advanced spectral methods for climatic time series. Rev Geophys 40(1):1–41

25. Golyandina N (2010) On the choice of parameters in singular spectrum analysis and related subspace-based methods. Stat Interface 3(3):259–279

26. Golyandina N (2019) Statistical approach to detection of signals by Monte Carlo singular spectrum analysis: multiple testing. arXiv:1903.01485

27. Golyandina N (2020) Particularities and commonalities of singular spectrum analysis as a method of time series analysis and signal processing. WIREs Comput Stat 12(4):e1487

28. Golyandina N, Shlemov A (2017) Semi-nonparametric singular spectrum analysis with projection. Stat Interface 10(1):47–57

29. Golyandina N, Nekrutkin V, Zhigljavsky A (2001) Analysis of time series structure: SSA and related techniques. Chapman&Hall/CRC, Boca Raton

30. Golyandina N, Korobeynikov A, Shlemov A, Usevich K (2015) Multivariate and 2D extensions of singular spectrum analysis with the Rssa package. J Stat Softw 67(2):1–78

31. Golyandina N, Korobeynikov A, Zhigljavsky A (2018) Singular spectrum analysis with R. Springer, Berlin

32. Greco G, Rosa R, Beskin G, Karpov S, Romano L, Guarnieri A, Bartolini C, Bedogni R (2011) Evidence of deterministic components in the apparent randomness of GRBs: clues of a chaotic dynamic. Sci Rep 1:91

33. Grenander U, Szegö G (1984) Toeplitz forms and their applications. Chelsea, New York

34. Groth A, Ghil M (2015) Monte Carlo singular spectrum analysis (SSA) revisited: detecting oscillator clusters in multivariate datasets. J Clim 28(19):7873–7893

35. Gruszczynska M, Klos A, Gruszczynski M, Bogusz J (2016) Investigation of time-changeable seasonal components in the GPS height time series: a case study for Central Europe. Acta Geodynamica et Geomaterialia 13(3):281–289

36. Hassani H, Heravi S, Zhigljavsky A (2009) Forecasting European industrial production with singular spectrum analysis. Int J Forecast 25(1):103–118

37. Hassani H, Xu Z, Zhigljavsky A (2011) Singular spectrum analysis based on the perturbation theory. Nonlinear Anal: Real World Appl 12(5):2752–2766

38. Hassani H, Heravi S, Zhigljavsky A (2013) Forecasting UK industrial production with multivariate singular spectrum analysis. J Forecast 32(5):395–408

39. Hassani H, Silva ES, Antonakakis N, Filis G, Gupta R (2017) Forecasting accuracy evaluation of tourist arrivals. Ann Tour Res 63:112–127

40. Hou Z, Wen G, Tang P, Cheng G (2014) Periodicity of carbon element distribution along casting direction in continuous-casting billet by using singular spectrum analysis. Metall Materials Trans B 45(5):1817–1826

41. Hsieh W, Wu A (2002) Nonlinear multichannel singular spectrum analysis of the tropical pacific climate variability using a neural network approach. J Geophys Res: Oceans 107(C7):13

42. Hua Y (1992) Estimating two-dimensional frequencies by matrix enhancement and matrix pencil. IEEE Trans Signal Process 40(9):2267–2280

43. Huang NE, Wu Z (2008) A review on Hilbert-Huang transform: Method and its applications to geophysical studies. Rev Geophys 46(2)

44. Itoh N, Marwan N (2013) An extended singular spectrum transformation (SST) for the investigation of Kenyan precipitation data. Nonlinear Process Geophys 20(4)

45. Jemwa GT, Aldrich C (2006) Classification of process dynamics with Monte Carlo singular spectrum analysis. Comput & Chem Eng 30(5):816–831

46. Keppenne C, Ghil M (1992) Adaptive filtering and prediction of the southern oscillation index. J Geophys Res: Atmos 97(D18):20,449–20,454

47. Khan MAR, Poskitt DS (2013) A note on window length selection in singular spectrum analysis. Austr & New Zealand J Stat 55(2):87–108

48. Kondrashov D, Ghil M (2006) Spatio-temporal filling of missing points in geophysical data sets. Nonlinear Process Geophys 13(2):151–159
49. Korobeynikov A (2010) Computation- and space-efficient implementation of SSA. Stat Interface 3(3):357–368
50. Korobeynikov A, Larsen RM, Wu KJ, Yamazaki I (2020) svd: Interfaces to various state-of-art SVD and eigensolvers. http://CRAN.R-project.org/package=svd, R package version 0.5
51. Korobeynikov A, Shlemov A, Usevich K, Golyandina N (2020) Rssa: a collection of methods for singular spectrum analysis. http://CRAN.R-project.org/package=Rssa, R package version 1.0.2
52. Kühnel A, Bogner C (2017) In-situ prediction of soil organic carbon by vis-NIR spectroscopy: an efficient use of limited field data. Eur J Soil Sci 68(5):689–702
53. Leles M, Sansão J, Mozelli L, Guimarães H (2018) Improving reconstruction of time-series based in singular spectrum analysis: a segmentation approach. Digital Signal Process 77:63–76
54. Mahecha MD, Fürst LM, Gobron N, Lange H (2010) Identifying multiple spatiotemporal patterns: a refined view on terrestrial photosynthetic activity. Pattern Recogn Lett 31(14):2309–2317
55. Mees AI, Rapp PE, Jennings LS (1987) Singular-value decomposition and embedding dimension. Phys Rev A 36:340–346
56. Merino Del Pozo S, Standaert FX (2015) Blind source separation from single measurements using singular spectrum analysis. In: Güneysu T, Handschuh H (eds) Cryptographic hardware and embedded systems - CHES 2015. Springer, Berlin, pp 42–59
57. Mi X, Liu H, Li Y (2019) Wind speed prediction model using singular spectrum analysis, empirical mode decomposition and convolutional support vector machine. Energy Convers Manag 180:196–205
58. Mohammad Y, Nishida T (2011) On comparing SSA-based change point discovery algorithms. IEEE SII pp 938–945
59. Moskvina V, Zhigljavsky A (2003) An algorithm based on singular spectrum analysis for change-point detection. Commun Stat Simul Comput 32(2):319–352
60. Nekrutkin V (2010) Perturbation expansions of signal subspaces for long signals. Stat Interface 3:297–319
61. Nilsson M (2004) Singular spectrum time-series analysis and continuous transformation groups. Proc R Soc Lon Ser A: Math Phys Eng Sci 460(2043):929–938
62. Noonan J, Zhigljavsky A (2018) Approximations of the boundary crossing probabilities for the maximum of moving weighted sums. Stat Papers 59(4):1325–1337
63. Okhotnikov G, Golyandina N (2019) EOP time series prediction using singular spectrum analysis. In: Corpetti T, Ienco D, Interdonato R, et al (eds) Proceedings of MACLEAN: MACHine learning for EArth observation workshop, RWTH Aahen University, CEUR Workshop Proceedings
64. Palus M, Novotná D (2004) Enhanced Monte Carlo singular system analysis and detection of period 7.8 years oscillatory modes in the monthly NAO index and temperature records. Nonlinear Process Geophys 11(5/6):721–729
65. Patterson K, Hassani H, Heravi S, Zhigljavsky A (2011) Multivariate singular spectrum analysis for forecasting revisions to real-time data. J Appl Stat 38(10):2183–2211
66. de Prony G (1795) Essai expérimental et analytique sur les lois de la dilatabilité des fluides élastiques et sur celles de la force expansive de la vapeur de l'eau et la vapeur de l'alkool à différentes températures. J de l'Ecole Polytechnique 1(2):24–76
67. Rekapalli R, Tiwari R, Dhanam K, Seshunarayana T (2014) Tx frequency filtering of high resolution seismic reflection data using singular spectral analysis. J Appl Geophys 105:180–184
68. Sanei S, Hassani H (2015) Singular spectrum analysis of biomedical signals. CRC Press
69. Santamaria I, Pantaleón C, Ibanez J (2000) A comparative study of high-accuracy frequency estimation methods. Mech Syst Signal Process 14(5):819–834
70. Sauer Y, Yorke J, Casdagli M (1991) Embedology. J Stat Phys 65:579–616

71. Shlemov A, Golyandina N (2014) Shaped extensions of Singular Spectrum Analysis. In: 21st international symposium on mathematical theory of networks and systems, July 7–11, 2014. Groningen, The Netherlands, pp 1813–1820

72. Silva I, Alonso H (2020) New developments in the forecasting of monthly overnight stays in the North Region of Portugal. J Appl Stat 1–14

73. Sippel S, Zscheischler J, Heimann M, Otto F, Peters J, Mahecha M (2015) Quantifying changes in climate variability and extremes: pitfalls and their overcoming. Geophys Res Lett 42(22):9990–9998

74. Stoica P, Soderstrom T (1991) Statistical analysis of MUSIC and subspace rotation estimates of sinusoidal frequencies. IEEE Trans Signal Process 39(8):1836–1847

75. Tang Tsz Yan V, Wee-Chung L, Hong Y (2010) Periodicity analysis of DNA microarray gene expression time series profiles in mouse segmentation clock data. Stat Interface 3(3):413–418

76. Thébault E, Vigneron P, Langlais B, Hulot G (2016) A Swarm lithospheric magnetic field model to SH degree 80. Earth Planets Space 68(1):1–13

77. Unnikrishnan P, Jothiprakash V (2015) Extraction of nonlinear rainfall trends using singular spectrum analysis. J Hydrol Eng 20(12):05015,007,1–15

78. Usevich K (2010) On signal and extraneous roots in Singular Spectrum Analysis. Stat Interface 3(3):281–295

79. Vautard R, Ghil M (1989) Singular spectrum analysis in nonlinear dynamics, with applications to paleoclimatic time series. Phys D 35:395–424

80. Vautard R, Yiou P, Ghil M (1992) Singular-spectrum analysis: a toolkit for short, noisy chaotic signals. Phys D 58:95–126

81. Wang F, Shen Y, Li W, Chen Q (2018) Singular spectrum analysis for heterogeneous time series by taking its formal errors into account. Acta Geodyn Geomater 4(192):395–403

82. Weare BC, Nasstrom JS (1982) Examples of extended empirical orthogonal function analyses. Mon Weather Rev 110(6):481–485

83. Ye Y, Cheng Y, He W, Hou M, Zhang Z (2016) Combining nonlinear adaptive filtering and signal decomposition for motion artifact removal in wearable photoplethysmography. IEEE Sens J 16(19):7133–7141

84. Yiou P, Sornette D, Ghil M (2000) Data-adaptive wavelets and multi-scale singular-spectrum analysis. Phys D: Nonlinear Phenom 142(3):254–290

85. Zabalza J, Ren J, Wang Z, Marshall S, Wang J (2014) Singular spectrum analysis for effective feature extraction in hyperspectral imaging. IEEE Geosci Remote Sens Lett 11(11):1886–1890

86. Zabalza J, Ren J, Zheng J, Han J, Zhao H, Li S, Marshall S (2015) Novel two-dimensional singular spectrum analysis for effective feature extraction and data classification in hyperspectral imaging. IEEE Trans Geosci Remote Sens 53(8):4418–4433

87. Zhigljavsky A (ed) (2010) Statistics and its interface (Special issue on the singular spectrum analysis in time series), vol 3. Guest Editor

Chapter 2
Basic SSA

2.1 The Main Algorithm

2.1.1 Description of the Algorithm

Consider a real-valued time series $\mathbb{X} = \mathbb{X}_N = (x_1, \ldots, x_N)$ of length N. Assume that $N > 2$ and $\mathbb{X}$ is a nonzero series; that is, there exists at least one i such that $x_i \neq 0$. Let L $(1 < L < N)$ be some integer called *the window length* and $K = N - L + 1$. Basic SSA is an algorithm of time series analysis described below and consisting of two complementary stages: decomposition and reconstruction.

2.1.1.1 First Stage: Decomposition

1st Step: Embedding
To perform the *embedding* we map the original time series into a sequence of lagged vectors of size L by forming $K = N - L + 1$ *lagged vectors*

$$X_i = (x_i, \ldots, x_{i+L-1})^{\mathrm{T}} \quad (1 \leq i \leq K)$$

of size L. If we need to emphasize the size (dimension) of the vectors X_i, then we shall call them *L-lagged vectors*.

The *L-trajectory matrix* (or simply *the trajectory matrix*) of the time series $\mathbb{X}$ is

$$\mathbf{X} = [X_1 : \ldots : X_K] = (x_{ij})_{i,j=1}^{L,K} = \begin{pmatrix} x_1 & x_2 & x_3 & \ldots & x_K \\ x_2 & x_3 & x_4 & \ldots & x_{K+1} \\ x_3 & x_4 & x_5 & \ldots & x_{K+2} \\ \vdots & \vdots & \vdots & \ddots & \vdots \\ x_L & x_{L+1} & x_{L+2} & \ldots & x_N \end{pmatrix}. \quad (2.1)$$

© The Author(s), under exclusive license to Springer-Verlag GmbH, DE, part of Springer Nature 2020
N. Golyandina and A. Zhigljavsky, *Singular Spectrum Analysis for Time Series*, SpringerBriefs in Statistics, https://doi.org/10.1007/978-3-662-62436-4_2

The lagged vectors X_i are the columns of the trajectory matrix $\mathbf{X}$. Both, the rows and columns of $\mathbf{X}$ are subseries of the original time series. The operator, which maps time series to their trajectory matrices, is called *embedding operator* and denoted by $\mathcal{T} = \mathcal{T}_L$.

The (i, j)-th element of the matrix $\mathbf{X}$ is $x_{ij} = x_{i+j-1}$ which yields that $\mathbf{X}$ has equal elements on the 'antidiagonals' $i + j = $ const. (Hence the trajectory matrix is a *Hankel matrix*.) Formula (2.1) defines a one-to-one correspondence between the trajectory matrix of size $L \times K$ and the time series.

2nd Step: Singular Value Decomposition (SVD)
At this step, we perform the singular value decomposition (SVD) of the trajectory matrix $\mathbf{X}$. Set $\mathbf{S} = \mathbf{X}\mathbf{X}^{\mathrm{T}}$ and denote by $\lambda_1, \ldots, \lambda_L$ the *eigenvalues* of $\mathbf{S}$ taken in the decreasing order of magnitude ($\lambda_1 \geq \ldots \geq \lambda_L \geq 0$) and by $U_1, \ldots, U_L$ the orthonormal system of the *eigenvectors* of the matrix $\mathbf{S}$ corresponding to these eigenvalues.

Set $d = \text{rank } \mathbf{X} = \max\{i, \text{ such that } \lambda_i > 0\}$ and $V_i = \mathbf{X}^{\mathrm{T}}U_i/\sqrt{\lambda_i}$ ($i = 1, \ldots, d$); note that in real-life time series we usually have $d = L^*$ with $L^* = \min(L, K)$. In this notation, the SVD of the trajectory matrix $\mathbf{X}$ can be written as

$$\mathbf{X} = \mathbf{X}_1 + \ldots + \mathbf{X}_d, \tag{2.2}$$

where $\mathbf{X}_i = \sqrt{\lambda_i}U_i V_i^{\mathrm{T}}$. The matrices $\mathbf{X}_i$ have rank 1; such matrices are sometimes called *elementary matrices*. The collection $(\sqrt{\lambda_i}, U_i, V_i)$ will be called i-th *eigentriple* (abbreviated as ET) of the SVD (2.2).

2.1.1.2   Second Stage: Reconstruction

3rd Step: Eigentriple Grouping
Once the expansion (2.2) is obtained, the grouping procedure partitions the set of indices $\{1, \ldots, d\}$ into m disjoint subsets $I_1, \ldots, I_m$.

Let $I = \{i_1, \ldots, i_p\}$. Then the resultant matrix $\mathbf{X}_I$ corresponding to the group I is defined as $\mathbf{X}_I = \mathbf{X}_{i_1} + \ldots + \mathbf{X}_{i_p}$. The resultant matrices are computed for the groups $I = I_1, \ldots, I_m$ and the expansion (2.2) leads to the decomposition

$$\mathbf{X} = \mathbf{X}_{I_1} + \ldots + \mathbf{X}_{I_m}. \tag{2.3}$$

The procedure of choosing the sets $I_1, \ldots, I_m$ is called *eigentriple grouping*. If $m = d$ and $I_j = \{j\}$, $j = 1, \ldots, d$, then the corresponding grouping is called *elementary*.

4th Step: Diagonal Averaging
At this step, we transform each matrix $\mathbf{X}_{I_j}$ of the grouped decomposition (2.3) into a new series of length N. Let $\mathbf{Y}$ be an $L \times K$ matrix with elements y_{ij}, $1 \leq i \leq L, 1 \leq j \leq K$. Set $L^* = \min(L, K)$, $K^* = \max(L, K)$ and $N = L + K - 1$. Let $y_{ij}^* = y_{ij}$ if $L < K$ and $y_{ij}^* = y_{ji}$ otherwise. By making the *diagonal averaging* we transfer the matrix $\mathbf{Y}$ into the series $(y_1, \ldots, y_N)$ using the formula

$$y_k = \begin{cases} \dfrac{1}{k} \displaystyle\sum_{m=1}^{k} y^*_{m,k-m+1} & \text{for } 1 \le k < L^*, \\[2ex] \dfrac{1}{L^*} \displaystyle\sum_{m=1}^{L^*} y^*_{m,k-m+1} & \text{for } L^* \le k \le K^*, \\[2ex] \dfrac{1}{N-k+1} \displaystyle\sum_{m=k-K^*+1}^{N-K^*+1} y^*_{m,k-m+1} & \text{for } K^* < k \le N. \end{cases} \qquad (2.4)$$

This corresponds to averaging the matrix elements over the 'antidiagonals' $i + j = k + 1$: the choice $k = 1$ gives $y_1 = y_{1,1}$, for $k = 2$ we have $y_2 = (y_{1,2} + y_{2,1})/2$, and so on. Note that if the matrix $\mathbf{Y}$ is the trajectory matrix of some series $(z_1, \ldots, z_N)$, then $y_i = z_i$ for all i.

Diagonal averaging (2.4) applied to a resultant matrix $\mathbf{X}_{I_k}$ produces a *reconstructed series* $\widetilde{\mathbb{X}}^{(k)} = (\widetilde{x}_1^{(k)}, \ldots, \widetilde{x}_N^{(k)})$. Therefore, the initial time series $(x_1, \ldots, x_N)$ is decomposed into a sum of m reconstructed series:

$$x_n = \sum_{k=1}^{m} \widetilde{x}_n^{(k)} \quad (n = 1, 2, \ldots, N). \qquad (2.5)$$

The reconstructed series produced by the elementary grouping will be called *elementary reconstructed series*.

Remark 2.1 The Basic SSA algorithm has a natural extension to the complex-valued time series, which is called Complex SSA. The only difference in the description of the algorithms is purely formal: the transpose sign should be replaced with the sign of complex conjugate; see Sect. 2.5.7 for details of application of Complex SSA.

2.1.2 Analysis of the Four Steps in Basic SSA

The formal description of the steps in Basic SSA requires some clarification. In this section we briefly discuss the meaning of the procedures involved.

2.1.2.1 Embedding

Embedding is a mapping that transfers a one-dimensional time series $\mathbb{X} = (x_1, \ldots, x_N)$ into the multidimensional series $X_1, \ldots, X_K$ with vectors $X_i = (x_i, \ldots, x_{i+L-1})^T \in \mathbb{R}^L$, where $K = N - L + 1$. The parameter defining the embedding is the *window length* L, an integer such that $2 \le L \le N - 1$. Note that the trajectory matrix (2.1) possesses an obvious symmetry property: the transposed matrix $\mathbf{X}^T$ is the trajectory matrix of the same series $(x_1, \ldots, x_N)$ with window length equal to K rather than L.

Embedding is a standard procedure in time series analysis, signal processing and the analysis of non-linear dynamical systems. For specialists in dynamical systems, a common technique is to obtain the empirical distribution of all the pairwise distances between the lagged vectors X_i and X_j and then calculate the so-called correlation dimension of the time series. This dimension is related to the fractal dimension of the attractor of the dynamical system that generates the time series; see Sect. 1.3.1 for further discussion of the dynamical system approach. In this approach, L must be relatively small and K must be very large (formally, $K \to \infty$). Similarly, in the so-called Hankel structured low-rank matrix approximation (HSLRA) considered in Sect. 3.8.4, the usual practice is to choose $L = r + 1$, where r is the guessed rank of the approximation matrix.

In SSA, the window length L should be sufficiently large. In particular, L has to be large enough so that each L-lagged vector incorporates an essential part of the behaviour of the initial time series $\mathbb{X} = (x_1, \ldots, x_N)$. The use of large values of L gives us a possibility of considering each L-lagged vector X_i as a separate time series and investigating the dynamics of certain characteristics for this collection of time series. See Sect. 2.4.3 for a discussion on the choice of L.

2.1.2.2 Singular Value Decomposition (SVD)

The SVD can be described in different terms and be used for different purposes. Let us start with general properties of the SVD which are important for SSA.

As was already mentioned, the SVD of an arbitrary nonzero $L \times K$ matrix $\mathbf{X} = [X_1 : \ldots : X_K]$ is a decomposition of $\mathbf{X}$ in the form

$$\mathbf{X} = \sum_{i=1}^{d} \sqrt{\lambda_i} U_i V_i^{\mathrm{T}}, \tag{2.6}$$

where λ_i ($i = 1, \ldots, L$) are eigenvalues of the matrix $\mathbf{S} = \mathbf{X}\mathbf{X}^{\mathrm{T}}$ arranged in order of decrease, $d = \max\{i, \text{ such that } \lambda_i > 0\} = \operatorname{rank}\mathbf{X}$, $\{U_1, \ldots, U_d\}$ is the corresponding orthonormal system of the eigenvectors of the matrix $\mathbf{S}$, and $V_i = \mathbf{X}^{\mathrm{T}}U_i/\sqrt{\lambda_i}$.

Standard SVD terminology calls $\sqrt{\lambda_i}$ the *singular values*; the U_i and V_i are the *left* and *right singular vectors* of the matrix $\mathbf{X}$, respectively. If we define $\mathbf{X}_i = \sqrt{\lambda_i}U_i V_i^{\mathrm{T}}$, then the representation (2.6) can be rewritten in the form (2.2), i.e. as the representation of $\mathbf{X}$ as a sum of elementary matrices $\mathbf{X}_i$.

If all eigenvalues have multiplicity one, then the expansion (2.2) is uniquely defined. Otherwise, if there is at least one eigenvalue with multiplicity larger than 1, then there is a freedom in the choice of the corresponding eigenvectors. We shall assume that the eigenvectors are somehow chosen and the choice is fixed.

The equality (2.6) shows that the SVD possesses the following property of symmetry: $V_1, \ldots, V_d$ form an orthonormal system of eigenvectors for the matrix $\mathbf{X}^{\mathrm{T}}\mathbf{X}$ corresponding to the eigenvalues $\lambda_1, \ldots, \lambda_d$. Note that the rows and columns of the trajectory matrix are subseries of the original time series. Therefore, the left and right singular vectors also have a temporal structure and hence can also be regarded as time series.

The SVD (2.2) possesses a number of optimal features. One of these features is the following property: among all matrices $\mathbf{X}^{(r)}$ of rank $r < d$, the matrix $\sum_{i=1}^{r} \mathbf{X}_i$ provides the best approximation to the trajectory matrix $\mathbf{X}$, so that $\|\mathbf{X} - \mathbf{X}^{(r)}\|_F$ is minimum.

Here and below the (*Frobenius*) *norm* of a matrix $\mathbf{Y}$ is $\|\mathbf{Y}\|_F = \sqrt{\langle \mathbf{Y}, \mathbf{Y} \rangle_F}$, where the inner product of two matrices $\mathbf{Y} = \{y_{ij}\}_{i,j=1}^{q,s}$ and $\mathbf{Z} = \{z_{ij}\}_{i,j=1}^{q,s}$ is defined as

$$\langle \mathbf{Y}, \mathbf{Z} \rangle_F = \mathrm{tr}(\mathbf{Y}^T \mathbf{Z}) = \sum_{i,j=1}^{q,s} y_{ij} z_{ij}.$$

For vectors, the Frobenius norm is the same as the conventional Euclidean norm.

Note that $\|\mathbf{X}\|_F^2 = \sum_{i=1}^{d} \lambda_i$ and $\lambda_i = \|\mathbf{X}_i\|_F^2$ for $i = 1, \ldots, d$. Thus, we shall consider the ratio $\lambda_i / \|\mathbf{X}\|_F^2$ as the characteristic of the contribution of the matrix $\mathbf{X}_i$ in the expansion (2.2). Consequently, $\sum_{i=1}^{r} \lambda_i / \|\mathbf{X}\|_F^2$, the sum of the first r ratios, is the characteristic of the optimal approximation of the trajectory matrix by the matrices of rank r or less. Moreover, if $\lambda_r \neq \lambda_{r+1}$ then $\sum_{i=r+1}^{d} \lambda_i$ is the squared distance between the trajectory matrix $\mathbf{X}$ and the set of $L \times K$ matrices of rank $\leq r$.

Let us now consider the trajectory matrix $\mathbf{X}$ as a sequence of L-lagged vectors. Denote by $\mathcal{X}^{(L)} \subset \mathbf{R}^L$ the linear space spanned by the vectors $X_1, \ldots, X_K$. We shall call this space the L-*trajectory space* (or, simply, *trajectory space*) of the time series $\mathbb{X}$. To emphasize the role of the time series $\mathbb{X}$, we use notation $\mathcal{X}^{(L)}(\mathbb{X})$ rather than $\mathcal{X}^{(L)}$. The equality (2.6) shows that $\mathcal{U} = \{U_1, \ldots, U_d\}$ is an orthonormal basis in the d-dimensional trajectory space $\mathcal{X}^{(L)}$.

Setting $Z_i = \sqrt{\lambda_i} V_i, i = 1, \ldots, d$, we can rewrite the expansion (2.6) in the form $\mathbf{X} = \sum_{i=1}^{d} U_i Z_i^T$, and for the lagged vectors X_j we have $X_j = \sum_{i=1}^{d} z_{ji} U_i$, where the z_{ji} are the components of the vector Z_i. This means that the vector Z_i is composed of the i-th components of lagged vectors X_j represented in the basis $\mathcal{U}$.

Let us now consider the transposed trajectory matrix $\mathbf{X}^T$. Introducing $Y_i = \sqrt{\lambda_i} U_i$ we obtain the expansion $\mathbf{X}^T = \sum_{i=1}^{d} V_i Y_i^T$, which corresponds to the representation of the sequence of K-lagged vectors in the orthonormal basis $\{V_1, \ldots, V_d\}$. Thus, the SVD gives rise to two dual geometrical descriptions of the trajectory matrix $\mathbf{X}$.

The optimal feature of the SVD considered above may be reformulated in the language of multivariate geometry for the L-lagged vectors as follows. Let $r < d$. Then among all r-dimensional subspaces $\mathcal{L}_r$ of $\mathbf{R}^L$, the subspace spanned by $U_1, \ldots, U_r$ approximates these vectors in the best way; that is, the minimum of $\sum_{i=1}^{K} \mathrm{dist}^2(X_i, \mathcal{L}_r)$ is attained at $\mathcal{L}_r = \mathrm{span}(U_1, \ldots, U_r)$. The ratio $\sum_{i=1}^{r} \lambda_i / \sum_{i=1}^{d} \lambda_i$ is the characteristic of the best r-dimensional approximation of the lagged vectors.

Another optimal feature relates to the properties of directions determined by the eigenvectors $U_1, \ldots, U_d$. Specifically, the first eigenvector U_1 determines the direction such that the variation of the projections of the lagged vectors onto this direction is maximum. Every subsequent eigenvector determines the direction that is orthogonal to all previous directions, and the variation of the projections of the lagged vectors onto this direction is also maximum. It is, therefore, natural to call the direction of i-th eigenvector U_i the i-th principal direction. Note that the elementary matrices $\mathbf{X}_i = U_i Z_i^{\mathrm{T}}$ are built up from the projections of the lagged vectors onto i-th directions.

This view on the SVD of the trajectory matrix composed of L-lagged vectors and an appeal to associations with principal component analysis lead us to the following terminology. We shall call the vector U_i the i-th (principal) eigenvector, the vectors V_i and $Z_i = \sqrt{\lambda_i} V_i$ will be called the i-th factor vector and the i-th principal component, respectively.

Remark 2.2 The SVD of the trajectory matrices used in Basic SSA is closely related to the Principal Component Analysis (PCA) in multivariate analysis and the Karhunen-Loève transform (KLT) in the analysis of stationary time series. However, the SVD approach in SSA uses the specificity of the Hankel structure of the trajectory matrix: indeed, the columns and rows of this matrix have the same temporal sense as all they are subseries of the original time series. This is not so in PCA. Links between SSA and the KLT are discussed in Sect. 1.3.1.

Remark 2.3 In general, any orthonormal basis $P_1, \dots, P_d$ of the trajectory space can be considered in place of the SVD-generated basis consisting of the eigenvectors $U_1, \dots, U_d$. In this case, the expansion (2.2) takes place with $\mathbf{X}_i = P_i Q_i^{\mathrm{T}}$, where $Q_i = \mathbf{X}^{\mathrm{T}} P_i$. One of the examples of alternative bases is the basis of eigenvectors of the autocovariance matrix in Toeplitz SSA, see Sect. 2.5.2.2. Other examples can be found among the methods of multivariate statistics such as Independent Component Analysis and Factor Analysis with rotation, see Sect. 2.5.3.

For further discussion concerning the use of other procedures in place of SVD, see Sect. 2.5.6.

2.1.2.3 Grouping

Let us now comment on the grouping step, which is the procedure of arranging the matrix terms $\mathbf{X}_i$ in (2.2). Assume $m = 2$, $I_1 = I = \{i_1 \dots, i_r\}$ and $I_2 = \{1, \dots, d\} \setminus I$, where $1 \leq i_1 < \dots < i_r \leq d$.

The purpose of the grouping step is the separation of additive components of time series. Let us introduce the very important concept of separability; this concept is thoroughly described in Sect. 2.3.3. Suppose that the time series $\mathbb{X}$ is a sum of two time series $\mathbb{X}^{(1)}$ and $\mathbb{X}^{(2)}$; that is, $x_i = x_i^{(1)} + x_i^{(2)}$ for $i = 1, \dots, N$. Let us fix the window length L and denote by $\mathbf{X}$, $\mathbf{X}^{(1)}$ and $\mathbf{X}^{(2)}$ the L-trajectory matrices of the time series $\mathbb{X}$, $\mathbb{X}^{(1)}$ and $\mathbb{X}^{(2)}$, respectively.

Consider an SVD (2.2) of the trajectory matrix $\mathbf{X}$. (Recall that if all eigenvalues have multiplicity one, then this expansion is unique.) We shall say that the time series $\mathbb{X}^{(1)}$ and $\mathbb{X}^{(2)}$ are (weakly) separable by the decomposition (2.2), if there exists a collection of indices $I \subset \{1, \ldots, d\}$ such that $\mathbf{X}^{(1)} = \sum_{i \in I} \mathbf{X}_i$ and consequently $\mathbf{X}^{(2)} = \sum_{i \notin I} \mathbf{X}_i$.

In the case of separability, the contribution of $\mathbf{X}^{(1)}$, the first component in the expansion $\mathbf{X} = \mathbf{X}^{(1)} + \mathbf{X}^{(2)}$, is naturally measured by the share of the corresponding eigenvalues: $\sum_{i \in I} \lambda_i \big/ \sum_{i=1}^{d} \lambda_i$.

The separation of the time series by the decomposition (2.2) can be looked at from different perspectives. Let us fix the set of indices $I = I_1$ and consider the corresponding resultant matrix $\mathbf{X}_{I_1}$. If this matrix, and therefore $\mathbf{X}_{I_2} = \mathbf{X} - \mathbf{X}_{I_1}$, are Hankel matrices, then they are necessarily the trajectory matrices of certain time series that are separable by the expansion (2.2).

Moreover, if the matrices $\mathbf{X}_{I_1}$ and $\mathbf{X}_{I_2}$ are close to some Hankel matrices, then there exist time series $\mathbb{X}^{(1)}$ and $\mathbb{X}^{(2)}$ such that $\mathbb{X} = \mathbb{X}^{(1)} + \mathbb{X}^{(2)}$ and the trajectory matrices of these time series are close to $\mathbf{X}_{I_1}$ and $\mathbf{X}_{I_2}$, respectively (the problem of finding these time series is discussed below). In this case, we shall say that the time series are *approximately separable*.

Therefore, the purpose of the grouping step (that is, the procedure of arranging the indices $1, \ldots, d$ into groups) is to find the groups $I_1, \ldots, I_m$ such that the matrices $\mathbf{X}_{I_1}, \ldots, \mathbf{X}_{I_m}$ satisfy (2.3) and are close to certain Hankel matrices.

Let us now look at the grouping step from the viewpoint of multivariate geometry. Let $\mathbf{X} = [X_1 : \ldots : X_K]$ be the trajectory matrix of a time series $\mathbb{X}$, $\mathbb{X} = \mathbb{X}^{(1)} + \mathbb{X}^{(2)}$, and the time series $\mathbb{X}^{(1)}$ and $\mathbb{X}^{(2)}$ are separable by the decomposition (2.2); this corresponds to splitting the index set $\{1, \ldots, d\}$ into I and $\{1, \ldots, d\} \setminus I$.

The expansion (2.3) with $m = 2$ means that $U_1, \ldots, U_d$, the basis in the trajectory space $\mathcal{X}^{(L)}$, is being split into two groups of basis vectors. This corresponds to the representation of $\mathcal{X}^{(L)}$ as a product of two orthogonal subspaces (*eigenspaces*) $\mathcal{X}^{(L,1)} = \mathrm{span}(U_i, i \in I)$ and $\mathcal{X}^{(L,2)} = \mathrm{span}(U_i, i \notin I)$ spanned by $U_i, i \in I$, and $U_i, i \notin I$, respectively.

Separability of two time series $\mathbb{X}^{(1)}$ and $\mathbb{X}^{(2)}$ means that the matrix $\mathbf{X}_I$, whose columns are the projections of the lagged vectors $X_1, \ldots, X_K$ onto the eigenspace $\mathcal{X}^{(L,1)}$, is exactly the trajectory matrix of the time series $\mathbb{X}^{(1)}$.

Despite the fact that several formal criteria for separability can be introduced, the whole procedure of splitting the terms into groups (i.e., the grouping step) is difficult to formalize completely. This procedure is based on the analysis of the singular vectors U_i, V_i and the eigenvalues λ_i in the SVD expansions (2.2) and (2.6). The principles and methods of identifying the SVD components for their inclusion into different groups are described in Sect. 2.4.

Since each matrix component of the SVD is completely determined by the corresponding eigentriple, we shall talk about grouping of the eigentriples rather than grouping of the elementary matrices $\mathbf{X}_i$.

Note also that the case of two series components ($m = 2$) considered above is often more sensibly regarded as the problem of separating out a single component rather than the problem of separation of two terms. In this case, we are interested in only one group of indices, namely I.

In the problems of signal processing, the time series $\mathbb{X}^{(1)}$ is interpreted as a signal. In these problems, we often choose $I_1 = \{1, \ldots, r\}$ for some r and call $\mathbb{X}^{(1)}$ the *signal subspace*.

2.1.2.4 Diagonal Averaging

If the components of the time series are separable and the indices have been split accordingly, then all matrices in the expansion (2.3) are Hankel matrices. We thus immediately obtain the decomposition (2.5) of the original time series: for all k and n, $\widetilde{x}_n^{(k)}$ is equal to the entries $x_{ij}^{(k)}$ along the antidiagonal $\{(i, j)$, such that $i + j = n + 1\}$ of the matrix $\mathbf{X}_{I_k}$.

In practice, however, this situation is not realistic. In the general case, no antidiagonal consists of equal elements. We thus need a formal procedure of transforming an arbitrary matrix into a Hankel matrix and thereby into a series. As such, we shall consider the procedure of *diagonal averaging*, which defines values of the time series $\widetilde{\mathbb{X}}^{(k)}$ as averages at the corresponding antidiagonals of the matrices $\mathbf{X}_{I_k}$.

It is convenient to represent the diagonal averaging step with the help of the *hankelization* operator $\boldsymbol{\varPi}_{\mathcal{H}}$. This operator acts on an arbitrary $L \times K$-matrix $\mathbf{Y} = (y_{ij})$ in the following way: for $A_s = \{(l, k) : l + k = s, 1 \le l \le L, 1 \le k \le K\}$ and $i + j = s$ the element $\widetilde{y}_{ij}$ of the matrix $\boldsymbol{\varPi}_{\mathcal{H}} Y$ is

$$\widetilde{y}_{ij} = \sum_{(l,k)\in A_s} y_{lk} \Big/ |A_s|,$$

where $|A_s|$ denotes the number of elements in the set A_s.

The hankelization is an optimal procedure in the sense that the matrix $\boldsymbol{\varPi}_{\mathcal{H}} \mathbf{Y}$ is closest to $\mathbf{Y}$ (with respect to the Frobenius matrix norm) among all Hankel matrices of the corresponding size [18, Proposition 6.3]. In its turn, the Hankel matrix $\boldsymbol{\varPi}_{\mathcal{H}} \mathbf{Y}$ defines the time series uniquely by relating the values in the antidiagonals to the values in the time series.

By applying the hankelization procedure to all matrix components of (2.3), we obtain another expansion:

$$\mathbf{X} = \widetilde{\mathbf{X}}_{I_1} + \ldots + \widetilde{\mathbf{X}}_{I_m}, \tag{2.7}$$

where $\widetilde{\mathbf{X}}_{I_l} = \boldsymbol{\varPi}_{\mathcal{H}} \mathbf{X}_{I_l}$.

A sensible grouping leads to the decomposition (2.3) where the resultant matrices $\mathbf{X}_{I_k}$ are almost Hankel ones. This corresponds to approximate separability and implies that the pairwise inner products of different matrices $\widetilde{\mathbf{X}}_{I_k}$ in (2.7) are small.

Since all matrices on the right-hand side of the expansion (2.7) are Hankel matrices, each matrix uniquely determines the time series $\widetilde{\mathbb{X}}^{(k)}$ and we thus obtain (2.5), the decomposition of the original time series.

Note that by linearity $\boldsymbol{\Pi}_{\mathcal{H}} \mathbf{X}_I = \sum_{i \in I} \boldsymbol{\Pi}_{\mathcal{H}} \mathbf{X}_i$ and hence the order in which the Grouping and the Diagonal Averaging steps appear in Basic SSA can be reversed.

The procedure of computing the time series $\widetilde{\mathbb{X}}^{(k)}$ (that is, building up the group I_k plus diagonal averaging of the matrix $\mathbf{X}_{I_k}$) will be called *reconstruction of a time series component* $\widetilde{\mathbb{X}}^{(k)}$ *by the eigentriples* with indices in I_k. In signal processing problems with $I_1 = \{1, \ldots, r\}$, we can say that the signal is reconstructed by the r leading eigentriples.

2.2 Potential of Basic SSA

In this section we start illustrating the main capabilities of Basic SSA. Note that terms such as 'trend', 'smoothing', 'signal', and 'noise' are used here in their informal, common-sense meaning and will be commented on later.

We refer to [21, Chap. 2] and [22] for more examples and the corresponding R codes.

2.2.1 Extraction of Trends and Smoothing

2.2.1.1 Trends of Different Resolution

The example 'Production' (crude oil, lease condensate, and natural gas plant liquids production, monthly data from January 1973 to September 1997, $N = 297$) shows the capability of Basic SSA in extraction of trends that have different resolutions. Despite the time series has a seasonal component (and the corresponding component can be extracted together with the trend component), for the moment we do not pay attention to periodicities.

Consider Fig. 2.1, where we use $L = 120$ and illustrate two alternatives in the trend resolution. In Fig. 2.1a we see that the leading eigentriple gives a general tendency of the time series. Figure 2.1b shows that the first three eigentriples describe the behaviour of the data more accurately and show not only the general decrease of production, but also its growth from the middle 70s to the middle 80s.

2.2.1.2 Smoothing

The time series 'Tree rings' (tree ring width, annual, from 42B.C. to 1970) were collected by R. Tosh and has the ID code ITRDB CA051 in International Tree Ring Data Bank (https://www.ncdc.noaa.gov/data-access/paleoclimatology-data/datasets/tree-ring). The time series looks like an autoregressive–moving-average (ARMA) process. If the ARMA-type model is accepted, then it is often meaningless to look for any trend or periodicity. Nevertheless, we can smooth the time series

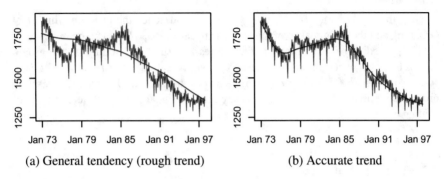

(a) General tendency (rough trend)   (b) Accurate trend

Fig. 2.1 Production: trends of different resolution

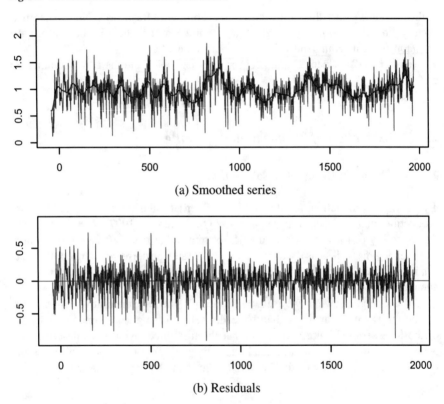

(a) Smoothed series

(b) Residuals

Fig. 2.2 Tree rings

with the help of Basic SSA. Figure 2.2a shows the initial time series and the result of its SSA smoothing, which is obtained by choosing the leading 3 eigentriples with window length $L = 100$. Figure 2.2b depicts the residuals.

Another example demonstrating SSA as a smoothing technique uses the 'White dwarf' data, which contains 618 point measurements of the time variation of the

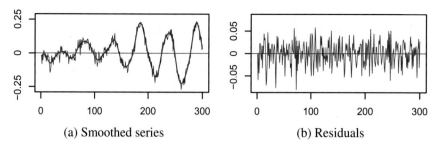

Fig. 2.3 White dwarf

intensity of the white dwarf star PG1159-035 during March 1989. The data is discussed in [8]. The whole series can be described as a smooth quasi-periodic curve with a noise component.

Using Basic SSA with window length $L = 100$ and choosing the leading 11 eigentriples for the reconstruction, we obtain the smooth curve of Fig. 2.3a (red thick line). The residuals (Fig. 2.3b) seem to have no evident structure (to simplify the visualization of the results these figures present only a part of the time series). Further analysis shows that the residual series can be regarded as a Gaussian white noise, though it does not contain very low frequencies. Thus, we can accept that in this case the smoothing procedure leads to noise reduction and the smooth curve in Fig. 2.3a describes the signal.

2.2.2   Extraction of Periodic Components

2.2.2.1   Extraction of Seasonality Components

Let us consider the extraction of seasonality components from the 'Production' data, which has been discussed in Sect. 2.2.1.1.

Again, choose $L = 120$. Simultaneously with trend we are able to extract seasonal components, gathering the harmonics produced by the fundamental period 12: 12 (ET19–20), 6 (ET15–16), 4 (ET9–10), 3 (ET13–15), 2.4 (ET4–5), and 2-months (ET7) harmonics. The resulting seasonal component is depicted in Fig. 2.4. This example demonstrates that SSA can perform seasonal adjustment even for time series with complex and changing seasonal behaviour.

2.2.2.2   Extraction of Cycles with Small and Large Periods

The time series 'Births' (number of daily births, Quebec, Canada, from January 1, 1977 to December 31, 1990) is discussed in [25]. It shows, in addition to a smooth

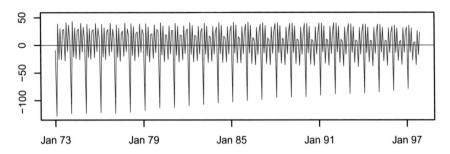

Fig. 2.4 Production: the seasonal component

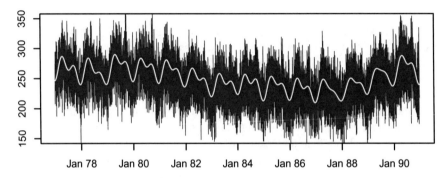

Fig. 2.5 Births: initial time series and its annual periodicity

trend, two cycles of different ranges: a one-year periodicity and a one-week periodicity.

Both periodicities (as well as the trend) can be simultaneously extracted by Basic
SSA with window length $L = 365$. Figure 2.5 shows the one-year cycle of the time
series added to the trend (white line) on the background of the 'Births' series from
1981 to 1990. Note that the form of this cycle varies in time, though the main
two peaks (spring and autumn) remain stable. The trend corresponds to the leading
eigentriple (ET1), while the one-year periodic component is reconstructed from ET
6–9. The eigentriples 12–19 also correspond to the fundamental period 365. However,
they are unstable in view of the small (with respect to the period value) window length.

Figure 2.6 demonstrates the one-week cycle on the background of the initial series
for the first four months of 1977. This cycle corresponds to ET 2–5 and ET 10–11.
The stability of the one-week periodicity does not seem to be related to the biological
aspects of the birth-rate.

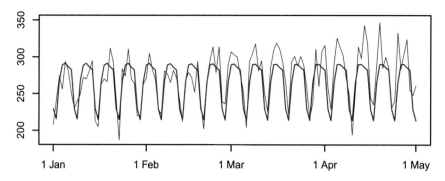

Fig. 2.6 Births: one-week periodicity

2.2.3 Complex Trends and Periodicities with Varying Amplitudes

The 'US unemployment' time series (unemployment of females (16–19 years) in thousands, US, monthly, from 1948 to 1981 [4]) serves as an example of SSA capability of extracting complex trends simultaneously with amplitude-modulated periodicities. The result of extraction is presented in Fig. 2.7a (the initial series and the reconstructed trend) and in Fig. 2.7b (seasonality).

The window length was taken as $L = 60$. Such a moderate window length was chosen in order to simplify the capture of the complex form of the trend and complex modulation of the seasonality. The trend is reconstructed from the ET 1, 8, 13 and 14 while the ET with numbers 2–7, 9–12 and 16 describe the seasonality.

2.2.4 Finding Structure in Short Time Series

The time series 'War' (U.S. combat deaths in the Indochina war, monthly, from 1966 to 1971 [31, Table 10]) is chosen to demonstrate the capability of SSA in finding a structure in short time series.

We have chosen $L = 18$. It is easy to see (Fig. 2.8a) that the two leading eigentriples perfectly describe the trend of the time series (the red thick line on the background of the initial data). This trend relates to the overall involvement of U.S. troops in the war.

Figure 2.8c shows the component of the initial series reconstructed from the ET 3–4. There is little doubt that this is an annual oscillation modulated by the war intensity. This oscillation has its origin in the climatic conditions: in South-East Asia, summer seasons are much harder for outdoor activities than winter seasons.

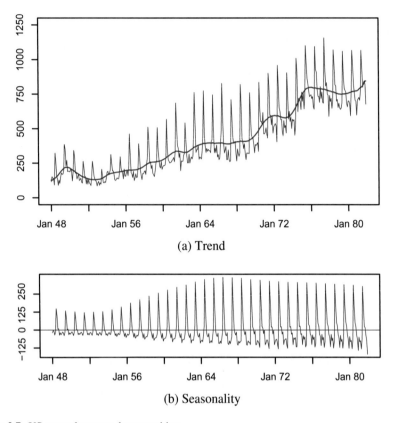

Fig. 2.7 US unemployment: decomposition

Two other series components, namely that of the quarterly cycle corresponding to the ET 5–6 (Fig. 2.8c) and the omitted 4-months cycle, which can be reconstructed from the ET 7–8, are both modulated by the war intensity and both are less clear for interpretation. Nevertheless, if we add all these effects together (that is, reconstruct the time series component corresponding to the eight leading eigentriples), a perfect agreement between the result and the initial series becomes apparent: see Fig. 2.8b with the red thick line corresponding to the reconstruction.

2.2.5 Envelopes of Oscillating Signals and Estimation of Volatility

The capabilities of SSA in separating signals with high and low frequencies can be used in a specific problem of enveloping highly oscillating sequences with slowly varying amplitudes.

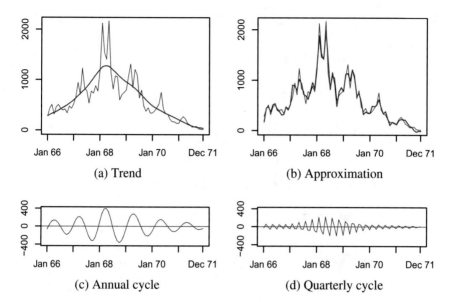

Fig. 2.8 War: structure of approximation

Let $x_n = A(n)\cos(2\pi\omega n)$, where the period $T = 1/\omega$ is not large in comparison with slowly varying $A(n)$. Define

$$y_n \stackrel{\text{def}}{=} 2x_n^2 = A^2(n) + A^2(n)\cos(4\pi\omega n). \tag{2.8}$$

Since $A^2(n)$ is slowly varying and the second term on the right-hand side of (2.8) oscillates rapidly, we can gather slowly varying terms of SSA decomposition for $(y_1, \ldots, y_N)$ and therefore approximately extract the term $A^2(n)$ from the time series (2.8). All we need to do then is to take the square root of the extracted term.

Let us illustrate this technique. Consider the square of the annual periodicity of the 'Germany unemployment' series (Fig. 2.32b in Sect. 2.5.4) multiplied by 2 and denote it by $\mathbb{Y}$. Taking window length $L = 36$ and reconstructing the low-frequency part of the time series $\mathbb{Y}$ from the eigentriples 1, 4, 7 and 10, we obtain an estimate of $A^2(n)$ (the reconstructed series are depicted in Fig. 2.9a by the red thick line; the blue thin line corresponds to the time series $\mathbb{Y}$). By taking the square root of the estimate we obtain the result (see Fig. 2.9b).

Very similarly we can analyze the dynamics of the variance of a heteroscedastic noise. Let $x_n = A(n)\varepsilon_n$, where ε_n is the white normal noise with zero mean and unit variance and $A(n)$ is a slowly changing function. Since $A^2(n) = \mathbf{D}x_n = \mathbf{E}x_n^2$, the trend extracted from the time series $\mathbb{Y}$ with $y_n = x_n^2$ provides an estimate of the variance.

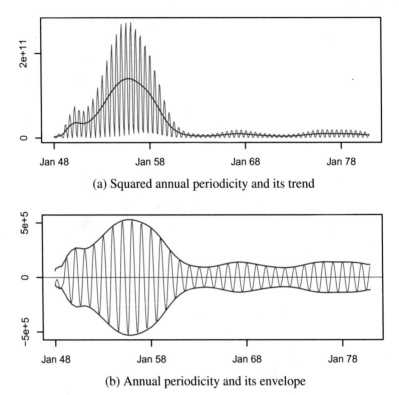

(a) Squared annual periodicity and its trend

(b) Annual periodicity and its envelope

Fig. 2.9 Germany unemployment: envelope construction

2.3 Models of Time Series and SSA Objectives

In the previous section, the terms 'trend', 'smoothing', 'amplitude modulation' and 'noise' were used without any explanation of their meaning. In this section, we shall provide related definitions and respective discussions. We shall also describe the major tasks that can be attempted by Basic SSA. Examples of application of Basic SSA for performing these tasks have been considered above in Sect. 2.2.

2.3.1 SSA and Models of Time Series

2.3.1.1 Models of Time Series and Periodograms

Formally, SSA can be applied to an arbitrary time series. However, a theoretical study of its properties requires specific considerations for different classes of series. Moreover, different classes assume different choices of parameters and expected results.

We thus start this section with description of several classes of time series, which are natural for SSA treatment, and use these classes to discuss the important concept of (approximate) separability defined earlier in Sect. 2.1.2.3. (For the theoretical aspects of separability, see [18].)

Since the main purpose of SSA is to make a decomposition of the time series into additive components, we always implicitly assume that this series is a sum of several simpler series. These 'simple' series are the objects of the discussion below. Note also that here we only consider deterministic time series, including those that can be regarded as 'noise'.

(a) Periodogram
For a description of the time series $\mathbb{X}_N = (x_1, \ldots, x_N)$ in terms of frequencies, it is convenient to use the language of the *Fourier expansion* of $\mathbb{X}_N$. This is the expansion

$$x_n = C_0 + \sum_{k=1}^{\lfloor N/2 \rfloor} \left(C_k \cos(2\pi n k/N) + S_k \sin(2\pi n k/N) \right), \tag{2.9}$$

where N is the length of the time series, $1 \le n \le N$, and $S_{N/2} = 0$ for even N. The constant C_0 is equal to the average of the time series, so that if the time series is centred, then $C_0 = 0$. Let $A_k^2 = C_k^2 + S_k^2$. Another form of (2.9) is

$$x_n = C_0 + \sum_{k=1}^{\lfloor N/2 \rfloor} A_k \cos(2\pi n k/N + \varphi_k).$$

We define the *periodogram* as

$$\Pi_x^N(k/N) = \begin{cases} C_0^2 & \text{for } k = 0, \\ (C_k^2 + S_k^2)/2 & \text{for } 0 < k < N/2, \\ C_{N/2}^2 & \text{for } k = N/2. \end{cases} \tag{2.10}$$

The last case in (2.10) is only possible if N is even. The normalization in the definition (2.10) is chosen to obtain

$$\|\mathbb{X}_N\|_F^2/N = \sum_{k=0}^{\lfloor N/2 \rfloor} \Pi_x^N(k/N). \tag{2.11}$$

Some other normalizations of the periodograms are known in literature and could be useful as well. The equality (2.11) implies that the value (2.10) of the periodogram at the point k/N describes the influence of the harmonic components with frequency $\omega = k/N$ within the sum (2.9).

The collection of frequencies $\omega_k = k/N$ with positive powers is called *support of the periodogram*. If the support of a certain periodogram belongs to some interval $[a, b]$, then this interval is called the *frequency range of the time series*.

Formally, the periodogram of the time series is an analogue of the spectral measure for stationary time series. Asymptotically, if the time series is stationary, then the periodograms approximate the spectral measures (see [18, Theorem 6.4]). The periodogram can also be helpful for a general description of an arbitrary time series. For example, trends can be classified as low-frequency time series.

The drawback of the periodogram analysis is its low resolution. In particular, periodograms can not distinguish frequencies that differ on any amount that is smaller than $1/N$. For short series, the grid $\{j/N, \; j = 0, \ldots, \lfloor N/2 \rfloor\}$ is a poor approximation to the whole range of frequencies $[0, 1/2]$, and the periodogram may not reveal a periodic structure of the time series components.

(b) Stationary Series

An infinite series (not necessarily stochastic) $\mathbb{X}_\infty = (x_1, x_2, \ldots, x_N, \ldots)$ is called *stationary* if for all nonnegative integers k, m we have

$$\frac{1}{N} \sum_{j=1}^{N} x_{j+k} x_{j+m} \xrightarrow[N \to \infty]{} R(k - m), \tag{2.12}$$

where the (even) function $R(\cdot)$ is called the *covariance function* of the time series $\mathbb{X}$ (the convergence in (2.12) is either deterministic or weak probabilistic depending on whether the time series is deterministic or stochastic). Below, when discussing stationarity, we shall always assume that $\frac{1}{N} \sum_{j=1}^{N} x_{j+k} \to 0$ (as $N \to \infty$) holds for any k, which is the zero-mean assumption for the original series.

The covariance function can be represented through the spectral measure, which determines properties of the corresponding stationary time series in many respects. The periodogram of a finite series $\mathbb{X}_N$ provides an estimate of the spectral density of $\mathbb{X}_\infty$.

A stationary time series $\mathbb{X}_\infty$ with discrete spectral measure m_x can normally be written as

$$x_n \sim \sum_{k} a_k \cos(2\pi \omega_k n) + \sum_{k} b_k \sin(2\pi \omega_k n), \quad \omega_k \in (0, 1/2], \tag{2.13}$$

where $a_k = a(\omega_k)$, $b_k = b(\omega_k)$, $b(1/2) = 0$ and the sum $\sum_k (a_k^2 + b_k^2)$ converges. (Note that $a(1/2) \neq 0$ if one of the ω_k is exactly $1/2$.) The form (2.13) for the time series $\mathbb{X}_\infty$ means that the measure m_x is concentrated at the points $\pm \omega_k$, (where $\omega_k \in (0, 1/2)$), with the weights $(a_k^2 + b_k^2)/4$. The weight of the point $1/2$ equals $a^2(1/2)$.

A series of the form (2.13) will be called *almost periodic*. *Periodic* series correspond to a spectral measure m_x concentrated at the points $\pm j/T$ $(j = 1, \ldots, \lfloor T/2 \rfloor)$ for some integer T. In terms of the representation (2.13), this means that the number of terms in this representation is finite and all frequencies ω_k are rational.

Almost periodic series that are not periodic are called *quasi-periodic*. For these series the spectral measure is discrete, but it is not concentrated on the nodes of any grid of the form $\pm j/T$. The *harmonic* $x_n = \cos(2\pi\omega n)$ with irrational ω provides an example of a quasi-periodic series.

Aperiodic (in other terminology, *chaotic*) series are characterized by a spectral measure that does not have atoms. In this case, one usually assumes the existence of the *spectral density*: $m_x(d\omega) = p_x(\omega)d\omega$. Aperiodic series are often used as models for *noise*. If the spectral density of an aperiodic stationary time series is constant, then this series is called *white noise*. Note that the white noise series does not have to be stochastic. In many cases, real-life stationary time series have both components, periodic (or quasi-periodic) and noise (aperiodic) components.

It is difficult, if not impossible, while dealing with finite series, to make a distinction between a periodic series with large period and a quasi-periodic series. Moreover, on finite time intervals aperiodic series are almost indistinguishable from a sum of harmonics with wide spectrum and small amplitudes.

(c) Amplitude-Modulated Periodicities

The definition of stationarity is asymptotic. This asymptotic nature has both advantages (for example, a rigorous mathematical definition allows an illustration of the main concepts by model examples) and disadvantages (for example, it is impossible to check the assumption of stationarity using a finite data set).

There are numerous deviations from stationarity. We consider only two classes of nonstationary time series which we describe at a qualitative level. Specifically, we consider amplitude-modulated periodic series and series with trends. The choice of these two classes is related to their practical significance and importance for SSA.

Trends are dealt with in the next subsection. Here we discuss the *amplitude-modulated* periodic signals; that is, series of the form $x_n = A(n)y_n$, where $\mathbb{Y} = (y_1, \ldots, y_N)$ is a periodic sequence and $A(n) \geq 0$. Usually it is assumed that on the given time interval ($n \in [1, N]$), the function $A(n)$ varies much slower than the low-frequency harmonic component of the time series $\mathbb{Y}$.

Series of this kind are typical in economics, where the period of the harmonics $\mathbb{Y}$ is related to seasonality, but the amplitude modulation is determined by long-term tendencies. Similar interpretation seems to be true for the example 'War', where the seasonal component of the combat deaths (Fig. 2.8c, d) is likely to be modulated by the intensity of the military activities.

Let us discuss the periodogram analysis of the amplitude-modulated periodic signals, temporarily restricting ourselves to the amplitude-modulated harmonic

$$x_n = A(n)\cos(2\pi\omega n + \theta), \quad n = 1, \ldots, N. \tag{2.14}$$

Unless the time series (2.14) is too short, its periodogram is supported on a short frequency interval containing ω. Indeed, for large $\omega_1 \approx \omega_2$ the sum

$$\cos(2\pi\omega_1 n) + \cos(2\pi\omega_2 n) = 2\cos\left(\pi(\omega_1 - \omega_2)n\right)\cos\left(\pi(\omega_1 + \omega_2)n\right)$$

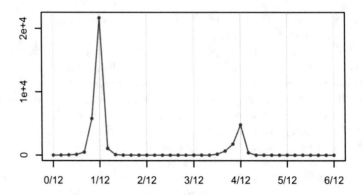

Fig. 2.10 War: periodogram of the main seasonality component

is a product of a slowly varying sequence $\{A(n)\}$, $A(n) = 2\cos\left(\pi(\omega_1 - \omega_2)n\right)$, and a harmonic with large frequency $(\omega_1 + \omega_2)/2$. The oscillatory nature of the function $A(n)$ cannot be seen for small N.

Figure 2.10 depicts the periodogram of the main seasonal (annual plus quarterly) component of the time series 'War' (Sect. 2.2.4). We can see that the periodogram is supported at around two main seasonal frequencies. However, it is not totally concentrated at these two points; this is caused by the amplitude modulation.

The discussion above implies that the appearance of exactly the same modulation can be caused by two different reasons: either it is the 'true' modulation or the modulation is spurious and originates from the closeness of the frequencies of the harmonic components of the original series.

Another reason for several frequencies to be spread around the main frequency is the discreteness of the periodogram grid $\{k/N\}$: if the frequency ω of a harmonic does not belong to the grid, then it spreads around the grid giving large positive values to two or more frequencies on the grid points in a close neighbourhood of ω.

Note that since the length of the 'War' series is proportional to 12, the frequencies $\omega = 1/12$ and $\omega = 1/3$, which correspond to annual and quarterly periodicities, fall exactly on the periodogram grid $\{k/36,\ k = 1, \ldots, 18\}$.

It is evident that not only periodic series can be modulated by the amplitude; the same can happen to the quasi-periodic and chaotic sequences. However, identification of these cases by means of the periodogram analysis is much more difficult.

(d) Trends

There is no commonly accepted definition of the concept 'trend'. Conventional approaches for defining trend either need postulating a parametric model (this would allow consideration of linear, exponential and logistic trends, among others) or consider a trend as a solution of an approximation problem, without any concerns about the tendencies; the most popular kind of trend approximation is the polynomial approximation.

In SSA framework, such meanings of the notion 'trend' are not suitable simply because Basic SSA is a model-free, and hence nonparametric, method. A generic definition of trend, appropriate for SSA, defines trend as an additive component of the time series which is (i) not stationary, and (ii) 'slowly varies' during the whole period of time where the time series was observed (compare [7, Chap. 2.12]).

Note that we have already collected oscillatory components of the time series into a separate class of (centred) stationary time series and therefore the term 'cyclical trend' does not make sense to SSA analysts.

Let us now discuss some consequences of this understanding of the notion 'trend'. The most important consequence is the issue of nonuniqueness of the solution to the problem 'trend identification' or 'trend extraction' in its nonparametric setup. This nonuniqueness has already been illustrated on the example 'Production', where Fig. 2.1 depicts two forms of the trend: a trend that explains a general tendency of the time series (Fig. 2.1a) and a detailed trend (Fig. 2.1b).

Furthermore, for finite time series, a harmonic component with very low frequency is practically indistinguishable from a trend (it can even be monotone on a finite time interval). In this case, supplementary subject-related information about the time series can be decisive for the problem of distinguishing trend from the periodicity. For instance, even though the reconstructed trend in the example 'War' (see Fig. 2.8a) looks like a periodicity observed over a time interval that is less than half of the period, there is no doubt that there is no periodicity in this case.

In the language of frequencies, any trend generates large powers in the region of low-frequencies in the periodogram. Moreover, we have assumed that any stationary time series is centred. Therefore, the average of all terms x_n of any series $\mathbb{X}$ is always added to its trend. On the periodogram, a nonzero constant component of the time series corresponds to an atom at zero.

(e) Additive Components of Time Series: Case Study
Summarizing, a general descriptive model of the time series that we use in SSA methodology is an additive model where the components of the time series are trends, oscillations and noise components. In addition, the oscillatory components are subdivided into periodic and quasi-periodic, while the noise components are aperiodic series. Amplitude modulation of the oscillatory and noise components is permitted. The sum of all additive components, except for the noise, will be referred to as *signal*.

Example 2.1 Let us consider the 'Rosé wine' time series (monthly rosé wine sales, Australia, from July 1980 to June 1994, thousands of litres). Figure 2.11 depicts the time series itself (blue thin line) and Fig. 2.12 presents its periodogram.

Figure 2.11 shows that the time series 'Rosé wine' has a decreasing trend and an annual periodicity of a complex form. Figure 2.12 shows the periodogram of the time series; it seems reasonable to assume that the trend is related to the large values at the low-frequency range, and the annual periodicity is related to the peaks at the frequencies $1/12$, $1/6$, $1/4$, $1/3$, $1/2.4$, and $1/2$. The non-regularity of powers of these frequencies indicates a complex form of the annual periodicity.

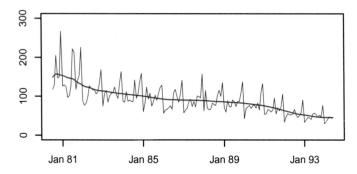

Fig. 2.11 Rosé wine: initial time series and the trend

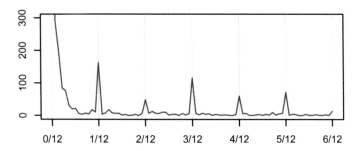

Fig. 2.12 Rosé wine: periodogram for the time series

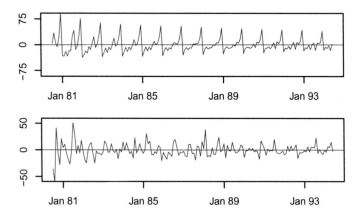

Fig. 2.13 Rosé wine: two components of the time series

We have applied Basic SSA with window length $L = 84$. Figure 2.13 depicts two additive components of the 'Rosé wine' series: the seasonal component (top graph), which is described by the ET 2–11, 13 and the residual series. The trend component (red thick line in Fig. 2.11) is reconstructed from the ET 1, 12, 14. Periodogram analysis demonstrates that the expansion of the time series into three parts is indeed

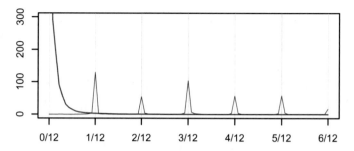

Fig. 2.14 Rosé wine: periodograms of the trend and the seasonal component

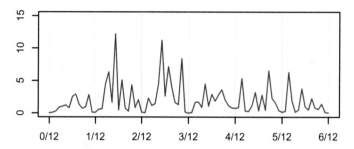

Fig. 2.15 Rosé wine: periodogram of the residuals

related to the separation of the spectral range into three regions: low frequencies correspond to the trend (red thick line in Fig. 2.14), the frequencies describing the seasonalities correspond to the periodic component (Fig. 2.14, blue thin line), and the residual series (which can be regarded as noise) has all the other frequencies (Fig. 2.15). Note that the periodograms of the whole series (see Fig. 2.12), its trend and the seasonal component (see Fig. 2.14) are presented on the same scale.   □

2.3.1.2 Models of Time Series and Rank

In the framework of SSA, the structure of a time series $\mathbb{X}$ is closely related to $d(L) =$ rank $\mathbf{X}$, the number of non-zero eigenvalues in the SVD of the trajectory matrix $\mathbf{X}$ (we shall call this number L-rank of the time series). If for some fixed d we have $d(L) = d$ for all large enough L, then the time series is called a finite-rank time series of rank d (see [18, Chap. 5] for details). For such series, we have $d(L) = \min(d, L)$ if $L \le K$.

For any time series of finite length, $d \le \min(L, K)$. If $d < \min(L, K)$, then the time series has a structure. A small value of d corresponds to a series with simple structure. If a time series component has small rank, then the grouping for its reconstruction is easier.

Let us consider several examples of time series models in terms of their rank. Note that the class of finite-rank time series includes sums of products of polynomials, exponentials and sinusoids.

Pure periodicities. Any sine-wave time series (so-called sinusoid) with frequency from the range $(0, 0.5)$ has rank 2, the saw-tooth sinusoid with frequency 0.5 has rank 1. Therefore, any almost periodic time series in the form (2.13) with finite number of addends has finite rank. Clearly, any periodic time series has finite rank. Aperiodic time series cannot have a finite rank.

Note that the simplicity of the sinusoid in the framework of SSA analysis depends on the number of the observed periods, while the fact that the rank of the sinusoid is equal to 2 is valid for the sinusoid of any frequency from $(0, 0.5)$.

Modulated periodicities. Modulation of periodicities can complicate or even destroy SSA structure of the time series. As a rule, for an arbitrary modulation, the modulated sinusoid is not of finite rank. The cosine modulation $A(n)$ defined in (2.14) is an example where the rank increases from 2 to 4 but stays finite.

The only possible example of modulation that does not change the rank of the signal is the exponential modulation $A(n) = \exp(\alpha n) = \rho^n$ with $\rho = e^\alpha$. For example, the rank of an exponentially damped sinusoid is the same as that of the undamped sinusoid. This is the essential advantage of SSA relative to the standard methods like the Fourier analysis and allows processing of the time series without log-transformation. Also, this allows SSA to deal with periodicities whose shape is changing.

Let us consider the 'Fortified wine' series (monthly fortified wine sales, Australia, from July 1980 to June 1994, thousands of litres). Figure 2.16 depicts the time series itself (blue thin line) and the reconstructed seasonality (red thick line); here the window length is $L = 84$ and the reconstruction is performed by ET 2–11. One can see that the form of seasonality is changing. This means that the standard methods of analysis like Fourier analysis cannot be used, even after the log-transformation. Figure 2.17 shows different kinds of modulation of the extracted (by Basic SSA) sine waves that altogether define the seasonal behaviour of the 'Fortified wine' series.

Trends. Trends have very different and, as a rule, non-structured behaviour; also, trends usually are the main contributors towards the non-stationarity of the time series. A typical trend (which is a slowly varying component of the time series) can be accurately approximated by a series of finite rank. The list of slowly-varying series with simple SSA structure and small rank includes an exponential series (rank 1), a sinusoid with large period (rank 2), a linear series (rank 2) and polynomials of higher order (rank > 2).

2.3.1.3 Additive and Multiplicative Models

By the definition, an additive model of a time series is a sum of components, while the multiplicative model is a product of positive components. Any multiplicative model can be easily transformed to an additive model by application of the log-transformation to the time series.

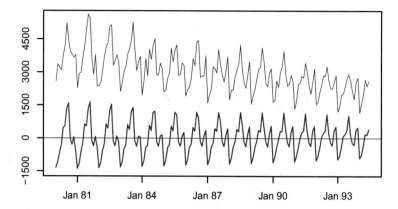

Fig. 2.16 Fortified wine: the initial time series and the reconstructed dynamic of the seasonality

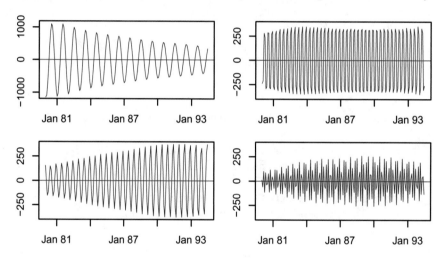

Fig. 2.17 Fortified wine: different behaviour of seasonal components

SSA deals with time series that can be represented as sums of components. One may think that SSA can not be used for time series represented via a multiplicative model. However, some series in a multiplicative model can be represented as sums with no extra transformation required. For example, let $x_n = t_n(1 + s_n)$, where $(t_1, \ldots, t_N)$ is a trend and $(s_1, \ldots, s_N)$ is a sinusoid with amplitude smaller than 1 (this is needed for positivity of $1 + s_n$). It is easily seen that $x_n = t_n + t_n s_n$; that is, the initial time series can be considered as a sum of a trend and a modulated sinusoid. Therefore, the multiplicative model can be considered as an additive one with modulated oscillations and noise.

Thus, SSA can be applied to both additive and multiplicative models. Log-transformation can increase the accuracy only if the structure of the signal after the log-transformation is simpler (has smaller rank) or the separability is improved.

Otherwise the log-transformation leads to a decrease of the accuracy of SSA analysis. For example, the log-transformation always worsens the structure of the time series with exponential trend.

2.3.1.4   Non-parametric Versus Parametric Models

To use Basic SSA we do not need to assume any model about the time series. Therefore, Basic SSA belongs to the class of nonparametric and model-free techniques. However, under the assumption of separability, a parametric model can be constructed based on SSA results. Let us demonstrate the idea.

Assume that the component $\mathbb{X}^{(1)}$ of the series $\mathbb{X} = \mathbb{X}^{(1)} + \mathbb{X}^{(2)}$ is L-separable and therefore has finite L-rank $r < \min(L, K)$. Let $\mathbb{X}^{(1)}$ be reconstructed by the r leading eigentriples, that is, $I_1 = \{1, \ldots, r\}$. Denote $\mathcal{X}^{(1)} = \mathrm{span}(U_1, \ldots, U_r)$ its trajectory space. If the L-th coordinate vector $\mathbf{e}_L = (0, \ldots, 0, 1)^\mathsf{T} \notin \mathcal{X}^{(1)}$, then $\mathbb{X}^{(1)}$ is governed by an LRR

$$x_n^{(1)} = \sum_{j=1}^{r} a_j x_{n-j}^{(1)}, \quad n = r+1, \ldots, N,$$

where the coefficients a_j are uniquely defined by the r-dimensional subspace $\mathcal{X}^{(1)}$, see [18, Chap. 5].

The coefficients a_j determine the complex numbers $\mu_1, \ldots, \mu_r$ which are the roots of the characteristic polynomial of the LRR, see Sect. 3.2 (we assume, for simplicity, that all roots μ_j are different; the case where some of μ_j are equal is more complicated and corresponds to the polynomial modulation of the time series components). The time series $\mathbb{X}^{(1)}$ can be written in terms of $\mu_1, \ldots, \mu_r$ as

$$x_n^{(1)} = \sum_{j=1}^{r} C_j \mu_j^n \tag{2.15}$$

with some coefficients C_j (see Theorem 3.1 in Sect. 3.2). Note that since $\mathbb{X}$ is a real-valued time series, if $\mu_j \in \{\mu_1, \ldots, \mu_r\}$ and μ_j is complex then there is the complex-conjugate $\mu_k = \mu_j^*$ of μ_j among $\{\mu_1, \ldots, \mu_r\}$. As we can write $\mu_j = \rho_j \exp(\mathrm{i}2\pi\omega_j)$, the set $\{\mu_j\}$ provides full information about the frequencies $\{\omega_j\}$. For known $\{\mu_j\}$, the coefficients C_j are determined by, for example, the values $x_1^{(1)}, \ldots, x_r^{(1)}$.

Since in practice there is no exact separability between time series components, many methods are developed to estimate coefficients of the parametric form of the time series component, see Sect. 3.8.

2.3.2 Classification of the Main SSA Tasks

Basic SSA can be very useful for solving the following problems of time series analysis: smoothing, extraction of trend and extraction of oscillatory components. The most general problem which Basic SSA may attempt to solve is that of finding the whole structure of the time series; that is splitting the time series into several 'simple' and 'interpretable' components, and the noise component. Let us discuss all these problems separately.

1. *Trend extraction and smoothing*
   There is no clear distinction between trend extraction and smoothing; for instance, the example 'US unemployment' (Fig. 2.7a) can be considered as an example of a refined trend extraction and as an example of smoothing. Neither of these two problems have exact meaning unless a parametric model is assumed. As a result, a large number of model-free methods can be applied to solve both of them. It is however convenient to distinguish between trend extraction and smoothing, at least on a qualitative level.
   Results of trend extraction by Basic SSA are illustrated on the examples 'Production' (Fig. 2.1a and b), 'US unemployment' (Fig. 2.7a) and 'War' (Fig. 2.8a). The example 'Tree rings' (Fig. 2.2a) shows smoothing capabilities of Basic SSA, see also [3, 19] (the latter demonstrates that smoothing by SSA could be used for non-parametric density estimation and can produce estimates that are more accurate than the celebrated kernel density estimates).
   Note that the problem of noise reduction is very similar to the problem of smoothing. The difference between these two problems is related to the conditions which the residual is expected to satisfy: for the noise reduction, the residual must not include any part of the signal whereas in the problem of smoothing the residual may include high-frequency periodic components.
2. *Extraction of oscillatory components*
   The general problem here is the identification and separation of oscillatory components of the time series that do not constitute parts of the trend. In the parametric form (under the assumptions of zero trend, finite number of harmonics, and additive stochastic white noise), this problem is extensively studied in the classical spectral analysis theory.
   Basic SSA is a model-free method. Therefore, the result of Basic SSA extraction of a single harmonic component of a time series is typically not a purely harmonic sequence. This is related to the fact that in practice we deal with an approximate separability rather than with the exact one (see Sect. 2.3.3).
   Basic SSA does not require assumptions about the number of harmonics and their frequencies. However, an auxiliary information about the initial time series always makes the situation clearer and helps in choosing parameters of the method, see Sect. 2.4.2.1.
   Finally, SSA allows the possibility of amplitude modulation for the oscillatory components of the time series. Examples 'War' (Sect. 2.2.4) and 'US unemploy-

ment' (Sect. 2.2.3) illustrate the capabilities of Basic SSA for the extraction of modulated oscillatory components.

3. *Splitting the series into 'simple' and 'interpretable' components and noise*
   This task can be thought of as a combination of two tasks considered above; specifically, the tasks of extraction of trend and extraction of periodic components. A specific feature of this task is that in the full decomposition the residual should consist of the noise only. Since model-free techniques often tend to find false interpretable components in noise, it is highly recommended to have a clear explanation (obtained using an information additional to the time series data itself) for each signal component found.

2.3.3 Separability of Components of Time Series

As discussed above, the main purpose of SSA is the decomposition of the original time series into a sum of series, so that each component in this sum can be identified as either a trend, periodic or quasi-periodic component (perhaps, amplitude-modulated), or noise. The notion of separability of series plays the fundamental role in the formalization of this problem (see Sects. 2.1.2.3 and 2.1.2.4). Roughly speaking, an SSA decomposition of the series $\mathbb{X}$ can be useful and informative only if the resulting additive components of the series are (approximately) separable from each other.

Weak and strong separability. Let us fix the window length L, consider a certain SVD of the L-trajectory matrix $\mathbf{X}$ of the initial series $\mathbb{X}$ of length N, and assume that the series $\mathbb{X}$ is a sum of two series $\mathbb{X}^{(1)}$ and $\mathbb{X}^{(2)}$, that is, $\mathbb{X} = \mathbb{X}^{(1)} + \mathbb{X}^{(2)}$. In this case, separability of the series $\mathbb{X}^{(1)}$ and $\mathbb{X}^{(2)}$ means (see Sect. 2.1.2.3) that we can split the matrix terms of the SVD of the trajectory matrix $\mathbf{X}$ into two different groups, so that the sums of terms within the groups give the trajectory matrices $\mathbf{X}^{(1)}$ and $\mathbf{X}^{(2)}$ of the series $\mathbb{X}^{(1)}$ and $\mathbb{X}^{(2)}$, respectively.

Since the SVD is not uniquely defined if there are multiple singular values, two types of separability can be considered. The separability is called *weak* if *there exists an SVD* of the trajectory matrix $\mathbf{X}$ such that we can split the SVD matrix terms into two different groups, so that the sums of terms within the groups give $\mathbf{X}^{(1)}$ and $\mathbf{X}^{(2)}$. The separability is called *strong*, if this is true *for any SVD* of the trajectory matrix.

For strong separability, it is necessary that the sets of eigenvalues produced by the SVDs of $\mathbf{X}^{(1)}$ and $\mathbf{X}^{(2)}$ have no intersection. Strong separability implies the weak one and is more desirable in practice. The absence of strong separability can be a serious problem for SSA. In Sect. 2.5.3.2 we introduce a method called SSA–ICA; this method (as well as DerivSSA described in [21, Sect. 2.5.3]) is able to improve strong separability in presence of (approximate) weak separability. Weak separability is easier to study and validate in practice. Although conditions for exact (weak) separability are rather restrictive, they can be extended to approximate separability and therefore be used in the practical analysis.

Each of the following conditions is equivalent to the definition of weak separability of two series $\mathbb{X}^{(1)}$ and $\mathbb{X}^{(2)}$:

1. any subseries of length L (and $K = N - L + 1$) of the series $\mathbb{X}^{(1)}$ is orthogonal to any subseries of the same length of the series $\mathbb{X}^{(2)}$ (the subseries of the time series are considered here as vectors); in term of trajectory matrices, $\mathbf{X}^{(1)}(\mathbf{X}^{(2)})^T = \mathbf{0}_{LL}$ and $(\mathbf{X}^{(1)})^T \mathbf{X}^{(2)} = \mathbf{0}_{KK}$;
2. the subspace $\mathcal{X}^{(L,1)}$ spanned by the columns of the trajectory matrix $\mathbf{X}^{(1)}$, is orthogonal to the subspace $\mathcal{X}^{(L,2)}$ spanned by the columns of the trajectory matrix $\mathbf{X}^{(2)}$, and similar orthogonality must hold for the subspaces $\mathcal{X}^{(K,1)}$ and $\mathcal{X}^{(K,2)}$ spanned by the rows of the trajectory matrices.

Characteristics of weak separability. Let $L^* = \min(L, K)$ and $K^* = \max (L, K)$. Introduce the weights

$$
w_i = \begin{cases} i & \text{for } 1 \le i < L^*, \\ L^* & \text{for } L^* \le i \le K^*, \\ N - i + 1 & \text{for } K^* < i \le N. \end{cases} \tag{2.16}
$$

The weight w_i in (2.16) is equal to the number of times the element x_i appears in the trajectory matrix $\mathbf{X}$ of the series $\mathbb{X} = (x_1, \ldots, x_N)$. Define the inner product of two series $\mathbb{X}^{(1)}$ and $\mathbb{X}^{(2)}$ of length N as

$$
\left(\mathbb{X}^{(1)}, \mathbb{X}^{(2)} \right)_w \overset{\text{def}}{=} \sum_{i=1}^{N} w_i x_i^{(1)} x_i^{(2)} \tag{2.17}
$$

and call the series $\mathbb{X}^{(1)}$ and $\mathbb{X}^{(2)}$ **w**-*orthogonal* if $\left(\mathbb{X}^{(1)}, \mathbb{X}^{(2)} \right)_w = 0$.

It follows from the separability conditions that separability implies **w**-orthogonality. To measure the degree of approximate separability between two series $\mathbb{X}^{(1)}$ and $\mathbb{X}^{(2)}$ we introduce the so-called **w**-correlation

$$
\rho^{(w)}(\mathbb{X}^{(1)}, \mathbb{X}^{(2)}) \overset{\text{def}}{=} \frac{\left(\mathbb{X}^{(1)}, \mathbb{X}^{(2)} \right)_w}{\|\mathbb{X}^{(1)}\|_w \|\mathbb{X}^{(2)}\|_w}. \tag{2.18}
$$

We shall loosely say that two series $\mathbb{X}^{(1)}$ and $\mathbb{X}^{(2)}$ are *approximately separable* if $\rho^{(w)}(\mathbb{X}^{(1)}, \mathbb{X}^{(2)}) \simeq 0$ for reasonable values of L (see [18, Sects. 1.5 and 6.1] for precise definitions). Note that the window length L enters the definitions of **w**-orthogonality and **w**-correlation, see (2.16).

Another qualitative measure of separability is related to the frequency structure of the time series components [18, Sect. 1.5.3]. It is sufficient (but not necessary) for weak separability that the supports of the periodograms of $\mathbb{X}^{(1)}$ and $\mathbb{X}^{(2)}$ do not intersect. If the intersection of the supports is, in a sense, small then the separability becomes approximate. Note that the separability of frequencies is equivalent to weak separability for the stationary time series.

Separable time series. Although there are many results available (see [18, Sects. 1.5 and 6.1]) on exact separability for the time series of finite rank, exact separability mostly presents purely theoretical interest. In practice, exact separability of components hardly ever occurs but an approximate separability can be achieved very often.

It is very important in practice that the trend, oscillations and noise components are approximately separable for large enough time series and window lengths.

To illustrate the concept of separability consider an example with two sinusoids

$$x_n^{(1)} = A_1 \cos(2\pi n\omega_1 + \varphi_1), \qquad x_n^{(2)} = A_2 \cos(2\pi n\omega_2 + \varphi_2), \qquad (2.19)$$

where $n = 1, \ldots, N, 0 < \omega_i < 0.5$ and $\omega_1 \neq \omega_2$. Let $L \leq N/2$ be the window length and $K = N - L + 1$. Let $L\omega_i$ and $K\omega_i$ be integers (in other words, let L and K be divisible by the periods $T_i = 1/\omega_i$, $i = 1, 2$). Then these sinusoids are weakly separable. The additional condition $A_1 \neq A_2$ implies strong separability, since the eigenvalues produced by the sinusoids are proportional to their squared amplitudes.

For large N and L two sinusoids are approximately weakly separable if $\omega_1 \neq \omega_2$; the divisibility of L and K by the periods is not necessary, although it can improve separability. The quality of separability (that influences the accuracy of the reconstruction) depends on the magnitude of $|\omega_1 - \omega_2|$. Close frequencies need much larger time series lengths to obtain a sufficient level of separability.

Under the condition of approximate weak separability, closeness of amplitudes A_1 and A_2 can cause the lack of strong separability. Note also that the frequency interpretation of separability for sinusoids is adequate, since for large L the leakage at the periodograms of sinusoids is small.

Theoretical approach to the asymptotic separability. The use of perturbation analysis is required to theoretically assess accuracy of SSA reconstruction in the case of asymptotic separability. Indeed, assume we observe a perturbed signal $\mathbb{X} = (x_1, \ldots, x_N) = \mathbb{S} + \delta\mathbb{P}$ of length N, where the time series $\delta\mathbb{P}$ is considered as a perturbation of the signal $\mathbb{S}$. The method results in $\widetilde{\mathbb{S}} = \mathbb{S} + \Delta_N(\mathbb{S}, \mathbb{P}, \delta)$; then $\Delta_N(\mathbb{S}, \mathbb{P}, \delta)$ is expanded as $\Delta_N(\mathbb{S}, \mathbb{P}, \delta) = \Delta_N^{(1)}(\mathbb{S}, \mathbb{P})\delta + \Delta_N^{(2)}(\mathbb{S}, \mathbb{P})\delta^2 + \ldots$. Most theoretical results assume that δ is small; hence, the first-order error $\Delta_N^{(1)}(\mathbb{S}, \mathbb{P})$ is studied [5, 24, 52]; this technique is in fact the linearization of the error in the neighbourhood of $\mathbb{S}$. For simplicity, the behavior of the first order error (as $\delta \to 0$) is frequently considered as $N \to \infty$. This technique allows one to obtain results, which partly help to understand the behaviour of the error. Nonetheless, this technique is still insufficient, since SSA works at any level of noise (that is, for any value of δ) as $N \to \infty$. In the series of papers [30, 39, 40], a step ahead in the study of perturbations without the assumption about the smallness of δ is made; the obtained results are related to the case of non-random perturbation and thereby to the separability of signal components.

It is also worth noting the existence of theoretical studies in the case of a fixed small L and stationary time series, where $\mathbf{XX}^{\mathrm{T}}/K$ tends to the $L \times L$ autocovariance matrix as N and $K = N - L + 1$ tend to infinity; see, e.g., [50]. Note, however, that the choice of small L is inappropriate from the viewpoint of separability.

2.4 Choice of Parameters in Basic SSA

In this section we discuss the role of parameters in Basic SSA and the principles for their selection. There are two parameters in Basic SSA: the first parameter is the window length L, and the second parameter is, loosely speaking, the way of grouping. In accordance with Sects. 2.3.1.1 and 2.3.2, we assume that the time series under consideration is a sum of a slowly varying trend, different oscillatory components, and a noise.

2.4.1 General Issues

2.4.1.1 Forms of Singular Vectors

We start with mentioning several theoretical results about the eigentriples of several time series with terms expressed in an analytic form.

Oscillations: exponential-cosine sequences. Consider the series

$$x_n = Ae^{\alpha n} \cos(2\pi \omega n + \varphi), \tag{2.20}$$

where $\omega \in (0, 1/2]$ and $\varphi \in [0, 2\pi)$. Depending on the values of parameters, the exponential-cosine sequence produces the following non-zero eigentriples:

1. *Exponentially modulated harmonic time series with frequency $\omega \in (0, 1/2)$*
   If $\omega \in (0, 1/2)$, then for any L and N the SVD of the trajectory matrix has two non-zero terms. Both eigenvectors (and factor vectors) have the form (2.20) with the same ω and α. In particular, for harmonic time series ($\alpha = 0$), the eigenvectors and factor vectors are harmonic series with frequency ω.
2. *Exponentially modulated saw-tooth curve ($\omega = 1/2$)*
   If $\sin(\varphi) \neq 0$, then x_n is proportional to $(-e^\alpha)^n$. If $\alpha = 0$, then $x_n = A(-1)^n = A\cos(\pi n)$. In this case, for any L the corresponding SVD has just one term. Both singular vectors have the same form as the initial series.

Let $\omega \neq 1/2$ and $\alpha = 0$. Then we have the pure harmonic time series defined by (2.20) with $\alpha = 0$. It generates the SVD with two non-zero terms; the singular vectors have the same harmonic form. Let us consider, for definiteness, the left singular vectors (that is, the eigenvectors) and assume an ideal situation, where $L\omega$ is integer. In this situation, the eigenvectors have the form of sine and cosine sequences with the same frequency ω and the same phases.

Figure 2.18 depicts pairwise scatterplots of four pairs of sin/cosine sequences with zero phases, the same amplitudes and frequencies $1/12$, $10/53$, $2/5$, and $5/12$. Clearly all the points lie on the unit circle. If $T = 1/\omega$ is an integer, then these points are the vertices of the regular T-vertex polygon. For the rational frequency

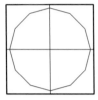 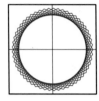 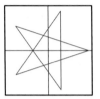 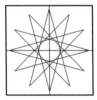

Fig. 2.18 Scatterplots of sines/cosines

$\omega = q/p < 1/2$ with relatively prime integers p and q, the points are the vertices of the regular p-vertex polygon.

Trends: exponential and polynomial series.

1. *The exponential sequence $x_n = e^{\alpha n}$.* For any N and window length L, the trajectory matrix of the exponential sequence has only one eigentriple. Both singular vectors of this eigentriple are exponential with the same parameter α.
2. *A general polynomial series.* Consider a polynomial series of degree m:

$$x_n = \sum_{k=0}^{m} A_k n^k, \quad A_m \neq 0.$$

For this series, the order of the corresponding SVD is $m + 1$ and all singular vectors are polynomials of degree not exceeding m.
3. *Linear series.* For a linear series $x_n = an + b, \quad a \neq 0$, with arbitrary N and L, the SVD of the L-trajectory matrix consists of two non-zero terms. The corresponding singular vectors are also linear series.

Note that the exponential-cosine and linear series (in addition to the sum of two exponential series with different rates) are the only series that have at most two non-zero terms in the SVD of their trajectory matrices for any series of length N and window length $L \geq 2$. This fact helps in their SSA identification as components of more complex series.

2.4.1.2 Predicting the Shape of Reconstructed Components

The shape of the eigentriples selected at the grouping stage can help us to predict the shape of the component which is going to be reconstructed from these eigentriples.

1. *If we reconstruct a component of a time series with the help of just one eigentriple and both singular vectors of this eigentriple have similar form, then the reconstructed component will have approximately the same form.* This means that when dealing with a single eigentriple we can often predict the behaviour of the corresponding component of the time series. For example, if both singular vectors of an eigentriple resemble linear series, then the corresponding component is also almost linear. If the singular vectors have the form of an exponential

series, then the trend has similar shape. Harmonic-like singular vectors produce harmonic-like components (compare this with the results for exponential-cosine series presented at the beginning of this section). This general rule also applies to some other properties of time series including monotonicity (monotone singular vectors generate monotone components of the series).

2. *If $L \ll K$ then the factor vector in the chosen eigentriple has a greater similarity with the component, reconstructed from this eigentriple, than the eigenvector.* Consequently, we can approximately predict the result of reconstruction from a single eigentriple by taking into account only the factor vector.

3. *If we reconstruct a series with the help of several eigentriples and the peri- odograms of their singular vectors are (approximately) supported on the same frequency interval $[a, b]$, then the frequency power of the reconstructed series will be mainly supported on $[a, b]$.* This feature is similar to that of item 1 but concerns several eigentriples and is formulated in terms of the Fourier expansions.

2.4.1.3 Eigenvalues

Let us detail several features of singular values of trajectory matrices.

1. The larger the singular value of the eigentriple is, the bigger the weight of the corresponding component of the time series. Roughly speaking, this weight may be considered as being proportional to the singular value.

2. By analogy with Principal Component Analysis (PCA), the share of the leading eigenvalues reflects the quality of approximation by the corresponding eigen- triples. However, there is a significant difference between Basic SSA and PCA, since PCA performs centering of variables. Since Basic SSA does not perform centering, the share of eigenvalues as a measure of approximation may have little sense. As an example, consider the series $\mathbb{X} = (x_1, \ldots, x_{100})$ with

$$x_n = c + \cos(2\pi n/10) + 0.9 \cos(2\pi n/5).$$

For $L = 50$ and $c > 0.45$ the three leading components provide exact recon- struction of $\mathbb{Y}$ with $y_n = c + \cos(2\pi n/10)$. It may be natural to suggest that the quality of approximation of $\mathbb{X}$ by $\mathbb{Y}$ should not depend on the value of c. How- ever, if we denote $p(c) = (\lambda_1 + \lambda_2 + \lambda_3)/(\lambda_1 + \ldots + \lambda_{50})$, then $p(0.5) \simeq 0.649$, $p(1) \simeq 0.787$ and $p(10) \simeq 0.996$.

3. For series $x_n = A \exp(\alpha n) \cos(2\pi \omega n)$, $\omega \in (0, 0.5)$, if $L\omega$ is integer, then both singular values coincide. If $\alpha \leq 0$ then for large N, L and $K = N - L + 1$, both singular values are close (formally, these values coincide asymptotically, as $L, K \to \infty$). Practically, they are close enough when L and K are several times larger than $T = 1/\omega$.

2.4.1.4 Elementary Reconstructed Components and w-Correlation Matrix

The elementary reconstructed series (recall that they correspond to the elementary grouping $I_j = \{j\}$) reflect the final result of reconstruction. If we group two eigen-triples, the i-th and j-th, then the reconstructed time series is equal to the sum of i-th and j-th elementary reconstructed components.

Let us use **w**–correlations as defined in Sect. 2.3.3 between elementary reconstructed components as separability measures.

While two singular vectors produced by a harmonic are orthogonal and have phase shift approximately equal to $\pi/2$, two associated elementary reconstructed series have approximately zero phase shift and therefore are strongly **w**-correlated. If two time series components are strongly separable, then the elementary reconstructed components produced by them are **w**-orthogonal. Therefore, the **w**-correlation matrix $\{\rho_{ij}^{(w)}\}$ between elementary reconstructed components reflects the structure of the time series detected by SSA.

The **w**-correlation matrix for an artificial series $\mathbb{X}$ with

$$x_n = e^{n/400} + \sin(2\pi n/17) + 0.5\sin(2\pi n/10) + \varepsilon_n, \quad n = 1, \ldots, 340, \quad (2.21)$$

standard Gaussian white noise ε_n, and $L = 85$, is depicted in Fig. 2.19, where **w**-correlations for the first 30 reconstructed components are shown in 20-colour scale from white to black corresponding to the absolute values of correlations from 0 to 1.

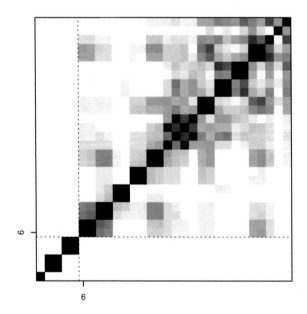

Fig. 2.19 Series (2.21): matrix of **w**-correlations

The leading eigentriple describes the exponential trend, the two pairs of the subsequent eigentriples correspond to the harmonics, and the large sparkling square indicates the white noise components. Note that this is in full agreement with the theory of (asymptotic) separability.

2.4.2 Grouping for Given Window Length

Assume that the window length L is fixed and we have already made the SVD of the trajectory matrix of the original time series. The next step is to group the SVD terms and hence to solve one of the problems discussed in Sect. 2.3.2. We suppose that the problem we want to solve has a solution; that is, the corresponding terms can be found in the SVD, and the result of the proper grouping would lead to the (approximate) separation of the time series components (see Sect. 2.3.3).

Therefore, we have to decide what the proper grouping is and how to construct it. In other words, we need to identify the eigentriples corresponding to the time series component we are interested in. Since each eigentriple consists of an eigenvector (left singular vector), a factor vector (right singular vector) and a singular value, this needs to be achieved using only the information contained in these vectors (considered as time series) and in the singular values.

2.4.2.1 Preliminary Analysis

The preliminary analysis of the time series is not necessary but it can be helpful for easier interpretation of the results of SSA processing. The following steps can be performed.

1. Observe the time series as a whole.

   - One can inspect the general shape of the trend, its complexity and hence how many trend components one can expect in the SVD expansion.
   - Based upon the form of the time series and its nature, one can expect some oscillations and their periods. For example, for seasonal monthly data, the period 12 is natural. If some period T is expected, then its divisors by integers (the result should be ≥ 2) are likely to be found in SSA decomposition. For monthly seasonal data they are $12, 6 = 12/2, 4 = 12/3, 3 = 12/4, 2.4 = 12/5$ and $2 = 12/6$.

2. Explore the time series periodogram.

   - Periodogram peaks reflect the expected periods that can be found in SSA decomposition.
   - Equal or close values at the peaks indicate a potential problem of the lack of strong separability.

For an example of a preliminary analysis of this kind, see the case study in Example 2.1 (Sect. 2.3.1.1), where Basic SSA was used to analyze the 'Rosé wine' series.

2.4.2.2 How to Group

For illustration, we provide references to the figures below in the description of the general recommendations. As an example, we consider the 'Fortified wine' time series (Fig. 2.16), which has already been analysed in Sect. 2.3.1.2.

General recommendations

1. Inspect the one-dimensional graphs of eigenvectors, factor vectors or elementary reconstructed components. Find slowly varying components. Note that any slowly varying component can be corrupted by oscillations if the trend and oscillating components are not separated. Elementary reconstructed components show whether the oscillating component is suppressed by the diagonal averaging. Most likely, the presence of the mix-up between the components is caused by the lack of strong separability. Changes in the window length and application of different preprocessing procedures can improve strong separability. All slowly varying components should be grouped into the trend group. Figure 2.20 shows the trend eigenvector and the trend reconstruction.
2. Consider two-dimensional plots of successive eigenvectors. Find regular p-vertex polygons, may be, in the form of a spiral. Group the found pairs of eigentriples. The harmonic with period 2 produces 1 eigentriple and therefore can be found at the one-dimensional graphs of, say, eigenvectors as a saw-tooth graph. See Fig. 2.21 with scatterplots and the reconstructed series in Fig. 2.17.
3. If there is a fundamental period T (e.g. seasonality with period 12), then special efforts should be made at finding the harmonics with periods that are divisors of T. Also, to reconstruct the whole periodic component with given period T, the pairs with periods T/k, $k = 1, \ldots, \lfloor T/2 \rfloor$ should be grouped, see Fig. 2.16, where the reconstruction of the whole seasonality is depicted.
4. The **w**-correlation matrix of elementary components can help in finding the components if they are not well separated and the techniques described above were

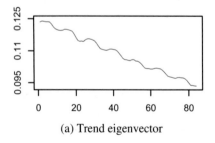

(a) Trend eigenvector

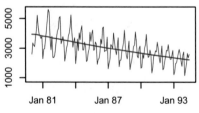

(b) Trend reconstruction and the initial series

Fig. 2.20 Fortified wine: trend component

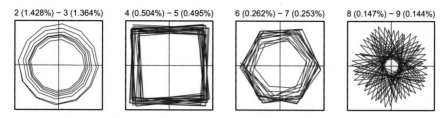

Fig. 2.21 Fortified wine: Scatterplots of eigenvector pairs corresponding to periods 12, 4, 6, 2.4

not successful. Blocks of two correlated components reflect a harmonic. A block of 4 correlated consequent components probably corresponds to two mixed pairs of harmonics. This can be checked by, for example, their periodogram analysis. Since noise is, in a sense, a mixture of many not-separable components, the **w**-correlation matrix can help to determine the number of components to identify.

2.4.2.3 How to Check the Results of Reconstruction

1. Any statistical testing is only possible when some assumptions are made. It could be a parametric model of the signal and noise. Nonparametric models usually require availability of a sample taken from the same distribution. The SSA procedure positions itself as a model-free technique and therefore justification of the results is complicated. Hence, interpretability of the resultant series components is very important. For example, extraction of a component with period 7 for monthly data is often more doubtful than, for example, half-year periodicity.
2. While signals could have very different forms and structures, noise frequently looks like white or rarer red noise. If there are reasons to assume a model of noise, then one can routinely test the corresponding hypothesis to confirm the results. In any case, the periodogram of the residual or their autocorrelation function can show if there is a part of the signal in the residual.
3. To test the specific hypothesis that the time series is red noise, Monte Carlo SSA (see the related part in Sect. 1.3) may be used.

2.4.2.4 Methods of Period Estimation

Since period estimation can be very useful in the process of identification of periodic components, let us discuss several methods of estimation that can be applied within the framework of SSA.

1. A conventional method for frequency estimation is periodogram analysis. We can apply it for estimation of frequencies presented in eigenvectors, factor vectors and reconstructed components. This can be effective for long series (and for large

window lengths if we want to consider eigenvectors). If the time series is short, then the resolution of the periodogram analysis is low.

2. We can estimate the period using both eigenvectors (or factor vectors) produced by a harmonic. If the eigenvectors have already been calculated, this method is very fast. Consider two eigentriples, which approximately describe a harmonic component with frequency $0 < \omega_0 < 0.5$. Then the scatterplot of their eigenvectors can be expressed as a two-dimensional curve with Euclidean components of the form

$$x(n) = r(n)\cos(2\pi\,\omega(n)\,n + \varphi(n)), \quad y(n) = r(n)\sin(2\pi\,\omega(n)\,n + \varphi(n)),$$

where the functions r, ω and φ are close to constants and $\omega(n) \approx \omega_0$. The polar coordinates of the curve vertices are $(r(n), \delta(n))$ with $\delta(n) = 2\pi\,\omega(n)\,n + \varphi(n)$. Since $\Delta_n \stackrel{\text{def}}{=} \delta(n+1) - \delta(n) \approx 2\pi\omega_0$, one can estimate ω_0 by averaging polar angle increments Δ_n ($n = 1, \ldots, L$). The same procedure can be applied to a pair of factor vectors.

3. We can also use the subspace-based methods of signal processing including ESPRIT, MUSIC, and others, see Sect. 3.8. These methods have high resolution and can be applied to short time series if we have managed to separate signal from noise accurately enough. An important common feature of these methods is that they do not require the sinusoids to be separated from each other.

2.4.3 Window Length

The window length L is the main parameter of Basic SSA: its inadequate choice would imply that no grouping activity will lead to a good SSA decomposition.

There is no universal rule for the selection of the window length. The main difficulty here is caused by the fact that variations in L may influence both weak and strong separability features of SSA, i.e., both the orthogonality of the appropriate subseries of the original series and the closeness of the singular values. However, there are several general principles for the selection of the window length L that have certain theoretical and practical grounds. Let us discuss these principles.

2.4.3.1   General Principles

1. The SVDs of the trajectory matrices, corresponding to the window lengths L and $K = N - L + 1$, are equivalent (up to the symmetry: left singular vectors $\leftrightarrow$ right singular vectors). Therefore, we can always assume $L \leq N/2$.
2. Assuming $L \leq N/2$, the larger L is, the more detailed is the decomposition of the time series. The most detailed decomposition is achieved when $L \simeq N/2$ unless

the series has finite rank d, see Sect. 2.3.1.2. In this case, SSA decompositions with any L such that $d \le L \le N + 1 - d$ are equivalent.

3. Small window lengths act like smoothing linear filters of width $2L - 1$. For small L, the filter produced by the leading eigentriple is similar to the Bartlett filter with triangular coefficients (see Sect. 3.9.3).

4. The following are the effects related to weak separability.

   - As the results concerning weak separability of time series components are mostly asymptotic (when $L, K \to \infty$), in the majority of examples to achieve better (weak) separation one has to choose large window lengths. In other words, the use of small L could lead to a mix-up between components which otherwise would be interpretable. Unless two time series are deterministic and exactly separable, there is no convergence of the reconstruction error to zero if L is fixed and $K \to \infty$ (for details, see [12]).
   - If the window length L is relatively large, then the (weak) separation is stable with respect to small perturbations in L.
   - On the other hand, for specific series and tasks, some concrete recommendations can be given for the window length selection; these recommendations can be very useful for relatively small N (see Sect. 2.4.3.3 below).

5. It is hard to successfully overcome (only by varying L) the difficulty related to the closeness of singular values; that is, to the absence of strong separability when there is an approximate weak separability. Let us mention two general points related to the closeness of the singular values.

   - For the series with complex structure, too large values of L can lead to an undesirable decomposition of the series components of interest, which in turn may yield their mixing with other series components. This is an unpleasant possibility, especially since a significant reduction of L can lead to a poor quality of the (weak) separation.
   - Alternatively, sometimes in these situations even a small variation in the value of L can reduce mixing and lead to a better separation of the components and hence provide a transition from weak to strong separability.

6. Whatever the circumstances, it is always a good idea to repeat SSA analysis several times using different values of L.

2.4.3.2 Window Length for Extraction of Trends and Smoothing

Trends

In the problem of trend extraction, a possible contradiction between the requirements for weak and strong separability emerges most frequently.

Since trend is a relatively smooth curve, its separability from noise and oscillations requires large values of L. On the other hand, if the trend has a complex structure, then for very large L it can only be described using a substantial number of eigentriples with relatively small singular values. Some of these singular values could be close to those generated by oscillations and/or noise time series components.

This happens in the example 'Births', see Sect. 2.2.2.2, where the window length of order 1000 and more (the time series length is 5113) leads to the situation where the components of the trend are mixed up with the components of the annual and half-year periodicities (other aspects relating to the choice of the window length in this example are discussed below).

If the trend is simple and dominates the rest of the series, then the choice of L does not present any difficulty (that is, L can be taken from a wide range). Let $\mathbb{X} = \mathbb{X}^{(1)} + \mathbb{X}^{(2)}$, where $\mathbb{X}^{(1)}$ is a trend and $\mathbb{X}^{(2)}$ is the residual. The notion of 'simplicity' can be understood as follows:

- From the theoretical viewpoint, the series $\mathbb{X}^{(1)}$ is well approximated by a series with finite and small rank d, see Sect. 2.3.1.2 for a description of the series of finite rank.
- We are interested in the extraction of the general tendency of the series rather than of the refined trend.
- In terms of frequencies, the periodogram of the series $\mathbb{X}^{(1)}$ is concentrated in the domain of small frequencies.
- In terms of SSA decomposition, the few first eigentriples of the decomposition of the trajectory matrix of the series $\mathbb{X}^{(1)}$ are enough for a reasonably good approximation of it, even for large L.

Assume also that the series $\mathbb{X}^{(1)}$ is much 'larger' than the series $\mathbb{X}^{(2)}$ (for instance, the inequality $\|\mathbb{X}^{(1)}\|_F \gg \|\mathbb{X}^{(2)}\|_F$ is valid).

Suppose that these assumptions hold and the window length L provides a certain (weak, approximate) separation between the time series $\mathbb{X}^{(1)}$ and $\mathbb{X}^{(2)}$. Then we can expect that in the SVD of the trajectory matrix of the series $\mathbb{X}$ the leading eigentriples will correspond to the trend $\mathbb{X}^{(1)}$; i.e., they will have larger singular values than the eigentriples corresponding to $\mathbb{X}^{(2)}$. In other words, we expect strong separability to occur. Moreover, the window length L, sufficient for the separation, should not be very large in this case in view of the 'simplicity' of the trend.

This situation is illustrated by the example 'Production' (Fig. 2.1a and b), where both trend versions are described by the leading eigentriples. However, more refined versions of the trend can be difficult to extract.

Much more complicated situations arise if we want to extract a refined trend $\mathbb{X}^{(1)}$, when the residual $\mathbb{X}^{(2)}$ has a complex structure (for example, it includes a large noise component) with $\|\mathbb{X}^{(2)}\|_F$ being large. Then large L can cause not only mixing of the ordinal numbers of the eigentriples corresponding to $\mathbb{X}^{(1)}$ and $\mathbb{X}^{(2)}$ (this is the case in the 'US unemployment' example), but also closeness of the corresponding singular values, and therefore the lack of strong separability.

Smoothing

The recommendations concerning the selection of the window length for the problem of smoothing are similar to those for trend extraction. This is related to the fact that

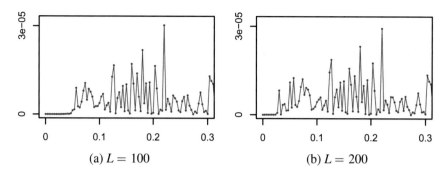

Fig. 2.22 White dwarf: periodograms of residuals

these two problems are closely related. Let us describe the effects of the window length in the language of frequencies.

If we treat smoothing as the removal of the high-frequency part of the time series, then we have to choose L large enough to provide separation of this low-frequency part from the high-frequency one. If the powers of all low frequencies of interest are significantly larger than those of the high ones, then the smoothing problem is not difficult, and our only job is to gather several leading eigentriples. This is the case for the 'Tree rings' and 'White dwarf' examples of Sect. 2.2.1.2. Here, the larger L we take, the narrower the interval of low frequencies we can extract.

For instance, in Sect. 2.2.1.2, the smoothing of the time series 'White dwarf' has been performed with $L = 100$, with the result of the smoothing being described by the leading 11 eigentriples. In the periodogram of the residuals (see Fig. 2.22a) we can see that for this window length the powers of the frequencies in the interval $[0, 0.05]$ are practically zero. If we take $L = 200$ and 16 leading eigentriples for the smoothing, then this frequency interval is reduced to $[0, 0.03]$ (see Fig. 2.22b). At the same time, for $L = 10$ and two leading eigentriples, the result of smoothing contains the frequencies from the interval $[0, 0.09]$.

Visual inspection shows that all smoothing results look similar. Also, their eigenvalue shares are equal to $95.9\% \pm 0.1\%$. Certainly, this effect can be explained by the following specific feature of the time series: its frequency power is highly concentrated in the narrow low-frequency region.

2.4.3.3 Window Length for Periodicities

The problem of choosing the window length L for extracting a periodic component $\mathbb{X}^{(1)}$ out of the sum $\mathbb{X} = \mathbb{X}^{(1)} + \mathbb{X}^{(2)}$ has certain specificities related to the correspondence between the window length and the period. These specificities are very similar for the pure harmonics, for complex periodicities and even for modulated periodicities. Thus, we do not consider these cases separately.

1. For the problem of extraction of a periodic component with period T, it is natural to measure the length of the series in terms of the number of periods: if $\mathbb{X}^{(1)}$ is asymptotically separable from $\mathbb{X}^{(2)}$, then to achieve the separation we must have, as a rule, the length of the series N such that the ratio N/T is at least several units.
2. For relatively short series, it is preferable to take into account the conditions for pure (nonasymptotic) separability (see Sect. 2.3.3); if one knows that the time series has a periodic component with integer period T (for example, $T = 12$), then it is advisable to take the window length L proportional to T. Note that from the theoretical viewpoint, $N-1$ must also be proportional to T.
3. In the case of long series, the requirement for L/T and $(N-1)/T$ to be integers is not that important if L is large enough (for instance, close to $N/2$) However, even in the case of long series it is recommended to choose L so that L/T is an integer.
4. If the series $\mathbb{X}^{(2)}$ contains a periodic component with period $T_1 \approx T$, then to extract $\mathbb{X}^{(1)}$ we generally need a larger window length than for the case when such a component is absent (see Sect. 2.3.3).

To demonstrate the effect of divisibility of L by T, let us consider the 'Eggs' example (eggs for a laying hen, monthly, U.S., from January 1938 to December 1940, [33, Chap. 45]). This time series has a relatively simple structure: it is the sum of an explicit annual oscillation (though not a harmonic one) and a trend, which is almost constant. However, this time series is short and therefore the choice of L is very important.

The choice $L = 12$ allows us to extract simultaneously all seasonal components (12, 6, 4, 3, 2.4, and 2-months harmonics) as well as the trend. The graph in Fig. 2.23 depicts the initial series and its trend (the red thick line), which is reconstructed from the leading eigentriple.

Figure 2.24a and b depict the matrices of **w**-correlations for the full decomposition of the series with $L = 12$ and $L = 18$. It is clearly seen that for $L = 12$ the matrix is essentially diagonal, which means that the eigentriples related to the trend and different seasonal harmonics are almost **w**-uncorrelated. This means that the choice $L = 12$ allows us to extract all harmonic components of the time series.

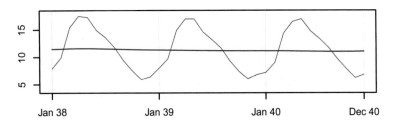

Fig. 2.23 Eggs: initial series and its trend

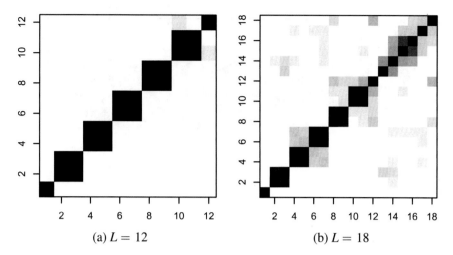

Fig. 2.24 Eggs: w-correlations

For $L = 18$ (that is, when the period 12 does not divide L), only the leading seasonality harmonics can be extracted properly.

The choice $L = 13$ would give results that are slightly worse than for $L = 12$, but much better than for $L = 18$. This confirms the robustness of the method with respect to small variations in L.

2.4.3.4 Refined Structure

In doing simultaneous extraction of different components from the whole time series, all the aspects discussed above should be taken into account. For instance, in basically all examples of Sect. 2.2, where the periodicities presented the main interest, the window length was a multiple of the periods. At the same time, if in addition trends were to be extracted, L was reasonably large (but smaller than $N/2$) to avoid the mix-up between the components.

To demonstrate the influence of the window length on the result of the decomposition, let us consider the example 'Births' (Sect. 2.2.2.2). In this series (daily data for about 14 years, $N = 5113$) there is a one-week periodicity ($T_1 = 7$) and an annual periodicity ($T_2 = 365$). Since $T_2 \gg T_1$, it is natural to take the window length as a multiple of T_2.

The choice $L = T_2$, as was shown in Sect. 2.2.2.2, guarantees a simultaneous extraction of both weekly and annual periodicities. Moreover, this window length also allows us to extract the trend of the series (see Fig. 2.25) using just one leading eigentriple. Note that the SSA reconstructions almost coincide for $L = 364$, 365 and 366.

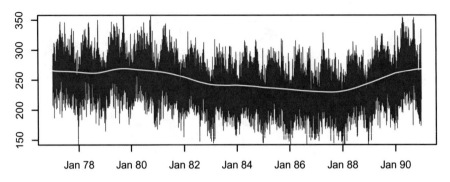

Fig. 2.25 Births: trend

At the same time, if we would choose $L = 3T_2 = 1095$ or $L = 7T_2 = 2555$, then the components of the trend will be mixed up with the components of the annual and half-year periodicities; this is a consequence of the complex shape of the trend and the closeness of the corresponding eigenvalues. Thus, choosing the values of L which are too large leads to the loss of strong separability.

If the problem of separation of the trend from the annual periodicity is not important, then values of L larger than 365 work well. If the window length is large, we can separate the global tendency of the time series (trend + annual periodicity) from the weekly periodicity + noise even better than for $L = 365$ (for $L = 1095$ this component is described by several dozen eigentriples rather than by 5 eigentriples for $L = 365$). In this case, the weekly periodicity itself is perfectly separable from the noise as well.

In even more complex cases, better results are often achieved by the application of the so-called Sequential SSA, see Sect. 2.5.4. In Sequential SSA, after extraction of a component with certain L, Basic SSA with different value of L is applied again, to the residual series obtained in the first run of SSA.

2.4.4   Signal Extraction

2.4.4.1   Condensed Form of SSA

For signal extraction, Basic SSA can be expressed by a concise formula. Consider the embedding operator $\mathcal{T}$, which transfers the time series into a Hankel matrix for a given window length L. Then all the steps of Basic SSA for constructing the estimate $\widetilde{\mathbb{S}}$ of the signal $\mathbb{S}$ from the time series $\mathbb{X}$ can be written as

$$\widetilde{\mathbb{S}} = \mathcal{T}^{-1} \circ \boldsymbol{\Pi}_{\mathcal{H}} \circ \boldsymbol{\Pi}_r \circ \mathcal{T}(\mathbb{X}), \tag{2.22}$$

where r is the number of the leading components chosen for reconstruction, Π_r is the projection of an $L \times K$ matrix into the space of $L \times K$ matrices of rank not larger than r, and $\Pi_{\mathcal{H}}$ is the hankelisation operator, which is the operator projecting any $L \times K$ matrix into the space of $L \times K$ Hankel matrices. In Basic SSA, both projections are performed with respect to the Frobenius norm.

2.4.4.2 Specifics of Extraction of the Signal

Sometimes, the structure of the deterministic component of the time series, which can be called a signal, is not important. In this case, the following three simple observations may help to achieve better separation of the signal from noise.

1. Since we are interested in the signal as a whole, separability of signal components is not essential. As a consequence, for the signal containing a periodic component, divisibility of the window length by the period is not important for separation of the signal from noise. However, if the window length is divisible by the period, it is easier to identify the signal components.
2. Since the signal components are often dominating, the only parameter of grouping is the number r of the leading components related to the signal. This number can be estimated using the matrix of w-correlations between elementary reconstructed components.
   In the example 'White dwarf' (Sect. 2.2.1.2) with $L = 100$, the matrix of the absolute values of w-correlations of the reconstructed components produced from the leading 30 eigentriples is depicted in Fig. 2.26 in the manner of Fig. 2.19. Splitting all eigentriples into two groups, from the first to the 11-th and the rest, gives a decomposition of the trajectory matrix into two almost orthogonal blocks, with the first block corresponding to the smoothed version of the original series and the second block corresponding to the residual, see Fig. 2.3a and b in Sect. 2.2.1.2
3. The problem of extraction of signal of finite rank from noisy time series is thoroughly studied. In particular, there are different methods of rank estimation (see below). These methods can be used while identifying the components in SSA.

2.4.4.3 Methods of Estimation of the Rank of the Signal

Two types of methods of rank estimation are used in signal processing. The first type is related to the so-called AIC-methods. They use some information criteria [53], which are based on the maximum likelihood function and therefore could only be applied to series of finite-rank with a given parametric model of the residuals (usually, Gaussian noise).

Consider a simple case of white Gaussian noise, where the ordinary LS estimate and the MLE coincide. Denote by $\widetilde{\mathbb{S}}(d)$ the LS estimate of the signal of length N assuming the parametric model of time series of rank d. The number of parameters is $2d$. Define $\mathrm{RSS}(d) = \|\widetilde{\mathbb{S}}(d) - \mathbb{S}\|^2$. Then, by the definition,

Fig. 2.26 White dwarf:
matrix of **w**-correlations

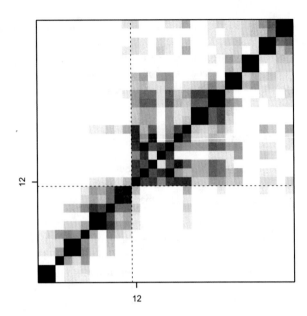

$$\text{AIC}(d) = N \ln(\text{RSS}(d)/N) + 4d, \qquad \text{BIC}(d) = N \ln(\text{RSS}(d)/N) + 2d \ln N.$$

The values of AIC/BIC can be used for choosing the rank r in a conventional way: the estimator of r is the minimizer of the considered information criterion. The main problem of this approach is related to its formulation as a parametric model; hence the approach may fail for real-life times series. Also, SSA does not provide the exact LS estimate; the LS estimate should be obtained by much more complex methods, see e.g. [57].

The second type of methods can be applied for general series. Let a method estimate certain time series characteristic. Then the accuracy of the estimation for different values of the assumed series rank r can point towards the proper value of r.

For example, the proper rank can be estimated on the base of the accuracy of forecasts of historical data. Or, more generally, one can consider several time series points as artificially missed values and their imputation accuracy serves as a characteristic for the choice of the best rank.

For signals of finite rank, specific methods can also be suggested. For example, the ESTER method is based on the features of the ESPRIT method as a method of parameter estimation (see for details Sect. 3.8.2).

2.4.5 Automatic Identification of SSA Components

While the choice of the window length is well supported by the SSA theory, the procedure for choosing the eigentriples for grouping is much less formal.

Let us describe several tests for the identification of SSA components constituting parts of the trend or related to periodicities. We assume that the components to be identified are (approximately) separated from the rest of the time series. The R-codes for automatic grouping can be found in [21, Sect. 2.7.3].

The tests described below can be used differently. First, these tests can provide some hints for making the grouping. This is a safe way of using the tests and we shall consider the tests from this viewpoint only. Second, the tests can be used as the base for the so-called *batch processing*. If there is a large set of similar time series, then a part of them can be used for the threshold adjustment. Similar to many other automatic procedures, the results of SSA batch processing may be misleading as many deviations from the model assumptions are possible. Note also that any choice of the threshold should take into consideration the following two conflicting types of decision error: (i) not to choose the proper SVD components, and (ii) to choose wrong SVD components. Certainly, to estimate probabilities of these errors, a stochastic model of the time series should be specified.

2.4.5.1   Grouping Based on w-correlations

The first approach is based on the properties of the **w**-correlation matrix $\{\rho_{ij}^{(w)}\}$ for separability identification, see Sect. 2.4.1.4. This has been used in different SSA-processing procedures. For example, Bilancia and Campobasso [6] consider hierarchical clustering with the dissimilarity measure $1 - |\rho_{ij}^{(w)}|$ and complete linkage, while Alonso and Salgado [2] use the k-means clustering procedure.

Let us consider two **w**-correlation matrices with full decompositions depicted in Figs. 2.19 and 2.26. The dissimilarity matrix consisting of $1 - |\rho_{ij}^{(w)}|$, along with the average linkage, provides the proper split into two clusters for the White dwarf data. The first cluster consists of ET1–11 and the second cluster corresponds to noise. The same procedure for the example of Fig. 2.19 gives the first cluster consisting of ET1 only, while the complete linkage provides the cluster of ET1–5. Note that the division into four groups (ET1; ET2,3; ET4,5; the rest) is the most appropriate for the average linkage. It seems that the average linkage is a good choice if the number of clusters is known. The choice of the number of clusters can be performed by the conventional tools of Cluster Analysis. Also, large **w**-correlations between grouped components from the clusters can help in identifying false clusters.

2.4.5.2 Identification of Trend

Since we define trend as any slowly-varying component of the time series, analysis of frequencies is a suitable tool for trend identification. The authors of [51] suggest using the number of zero crossings or the Kendall's test to find slowly-varying eigenvectors. A rather general approach is to use the periodogram and consider the contribution of low frequencies as a test; see e.g. [1], where the emphasis is made on the procedure of an automatic choice of the identification thresholds.

Consider the periodogram (2.10) of a series $\mathbb{Y}$ of length M and define

$$T(\mathbb{Y}; \omega_1, \omega_2) = \sum_{k:\omega_1 \leq k/M \leq \omega_2} I_y^M(k/M), \qquad (2.23)$$

where $I_y^M(k/M) = M \, \Pi_y^M(k/M)/\|\mathbb{Y}\|^2$, Π_y^M is defined in (2.10). In view of (2.11), $0 \leq T(\mathbb{Y}; \omega_1, \omega_2) \leq 1$ for any $0 \leq \omega_1 \leq \omega_2 \leq 0.5$. Let us choose the bounding frequency ω_0, $0 \leq \omega_0 \leq 0.5$, and set up a threshold T_0, $0 \leq T_0 \leq 1$.

Below we formulate a generic test for deciding whether a given SSA component is slowly varying. This test can be applied to eigenvectors, factor vectors and elementary reconstructed components considered as time series. Let $\mathbb{Y}$ be the series we are going to test.

Trend test T. *A given component $\mathbb{Y}$ is related to the trend if $T(\mathbb{Y}; 0, \omega_0) \geq T_0$.*

The choice of the bounding frequency ω_0 depends on how we want the trend to look like. For example, for monthly data with possible seasonality it is recommended to choose $\omega_0 < 1/12$.

If we consider the results of trend tests as hints for the eigentriple identification, it is not necessary to set the threshold value T_0, since we can simply consider the values of the test statistics $T(\mathbb{Y}; 0, \omega_0)$ for the series $\mathbb{Y}$ (the eigenvectors or the elementary reconstructed components) related to each eigentriple.

Let us consider the 'Production' example (Sect. 2.2.1.1, Fig. 2.1b), where a reasonably accurate trend is described by the three leading eigentriples. If we choose $\omega_0 = 1/24$ and $T_0 = 0.9$, then the described procedure identifies ET1–3,6,8,11,12; that is, the trend identified in this manner and depicted in Fig. 2.27 is even more accurate than that depicted in Fig. 2.1b. The result is stable with respect to the choice of the threshold and is exactly the same when we apply it to eigenvectors, factor vectors or reconstructed components. The values of the test $T(\,\cdot\,; 0, 1/24)$ applied to the 12 leading factor vectors are respectively: 0.9999, 0.9314, 0.9929, 0.0016, 0.0008, 0.9383, 0.0053, 0.9908, 0.0243, 0.0148, 0.9373, 0.9970. If we are interested in general tendency, then the test T with $\omega_0 = 1/120$ identifies one leading component only, the same result as in Fig. 2.1a.

For the 'Rosé wine' example, where the trend was extracted by ET1, 12, and 14, the test $T(\cdot; 0, 1/24)$ applied to 16 leading eigenvectors gives 0.9993 for ET1, 0.8684 for ET12, 0.9839 for ET14 and values smaller than 0.02 for all other eigentriples. This outcome perfectly agrees with visual examination.

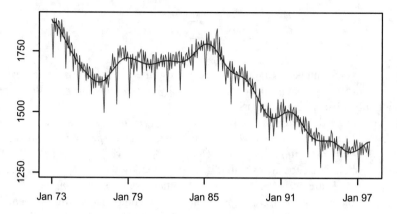

Fig. 2.27 Production: automatically identified refined trend

2.4.5.3 Identification of Harmonics

The method for identification of the harmonic components is based on the study of the corresponding singular vectors. Ideally, any harmonic component produces two eigenvectors, which are sine and cosine sequences if L and $K = N - L + 1$ are divisible by the period of the harmonic. Also, if $\min(L, K) \to \infty$ then the pair of the corresponding either left or right singular vectors tends to the sine and cosine sequences, correspondingly.

Define for $H, G \in \mathbb{R}^L$

$$\rho(G, H) = \max_{0 \le k \le L/2} \gamma(G, H; k), \text{ where } \gamma(G, H; k) = 0.5(I_g^L(k/L) + I_h^L(k/L))$$

and the quantity I is the same as in (2.23). It is clear that $\rho(G, H) \le 1$ and that for any integer $L\omega$ the equality $\rho(G, H) = 1$ is valid if and only if $h_n = \cos(2\pi \omega n + \varphi)$ and $g_n = \sin(2\pi \omega n + \varphi)$. Also, for arbitrary ω, $\rho(G, H) \to 1$ as $L \to \infty$.

Therefore, the value of $\rho(U_i, U_j)$ (as well as $\rho(V_i, V_j)$) can be used as an indicator of whether the pair of eigenvectors U_i, U_j (or factor vectors V_i, V_j) is produced by a harmonic component.

The case of amplitude-modulated harmonics is slightly more complicated. Let us consider the identification of the exponentially damped sine waves; recall that these waves are naturally generated by SSA. Both eigenvectors (and factor vectors) have the same form (2.20) with the same frequency ω and the exponential rate α. Therefore we principally can apply the $\rho(G, H)$ for their identification. However, the modulation leads to the decrease of $\rho(G, H)$ and this should be accounted for while choosing the threshold value.

Let us introduce a test which is a modification of the test suggested in [51] to take into consideration the leakage caused by possible modulation of the harmonics and location of their frequencies between positions in the periodogram grid. Define

$$\tau(G, H) = \max_{0 \leq k \leq L/2 - m_0} \sum_{j=0}^{m_0 - 1} \gamma(G, H; k + j),$$

where m_0 is some integer. Note that below we use the result stating that an exponentially damped sinusoid produces asymptotically equal eigenvalues. We therefore consider only adjacent eigenvectors.

Harmonic test τ. *An eigenvector pair* (U_j, U_{j+1}) *is identified as corresponding to some damped sinusoid if the periodograms of* U_j *and* U_{j+1} *are peaked at frequencies differing not more than* m_0/L *and* $\tau(U_j, U_{j+1}) \geq \tau_0$ *for given threshold* $\tau_0 \in [0, 1]$.

Here m_0 should be chosen equal to 0 if the period is known and we can choose L such that L and K are divisible by the period. Otherwise we choose $m_0 = 1$.

Note that the test above needs adjustments for the components with frequencies 0 and 0.5: the frequency 0 should not be considered as a candidate for periodicity, while the sine wave with frequency 0.5 is the saw-tooth function and produces just one component with frequency 0.5. Also, the test can be supplemented with the frequency estimation (see Sect. 2.4.2.4) and the results can be filtered in accordance with the chosen frequency range.

Let us apply the τ-test to the 'Production' example considered in Sects. 2.2.1.1 and 2.2.2.1. This time series has a trend of complex form and we need to set a period-based filter to distinguish between the cyclic components of the trend and the seasonal components. Assume that all possible periods fall into the interval [2,13]. Then the τ-test with thresholds τ_0 from the range 0.86–0.96 identifies the same seasonal components as were chosen in Sect. 2.2.2.1 by visual inspection except for the pair ET19–20 (period 12) with $\tau(U_{19}, U_{20}) = 0.85$. This is explained by the sharp decrease of the harmonic with period 12 and a poor separability of the annual harmonic.

Warning. Above we have considered examples with well separable components. However, if the separability is poor, then the automatic procedure may easily fail. This yields that automatic identification is useful for grouping but it can not replace the techniques that improve separability (see Sect. 2.5.3).

2.5 Some Variations of Basic SSA

A clever modification of the decomposition step in Basic SSA may visibly improve SSA accuracy. Note that modifications of the decomposition step have very little or no relation to the extensions of the embedding step of the SSA algorithm and therefore the modifications considered in this section can be applied to the multivariate and multidimensional extensions of SSA considered in Sect. 2.6.

Different modifications of the decomposition step will be considered in a common form $\mathbf{X} = \mathbf{X}_1 + \ldots + \mathbf{X}_d$, where the matrices $\mathbf{X}_i$ are some matrices of rank one. In Basic SSA, the SVD expansion into rank-one matrices is considered. If there is no additional information about the time series, the SVD is optimal. Several modifications of the decomposition step are related to different assumptions about the time series (Sects. 2.5.2.1 and 2.5.2.2). Other modifications of the SVD step supplement the SSA algorithm by an additional step, whose aim is to improve the decomposition of the signal subspace obtained from the SVD decomposition of the trajectory matrix. This nested decomposition can be performed by rotations of the basis $\{U_i\}_{i=1}^r$ of the estimated signal subspace obtained by the SVD (Sect. 2.5.3).

We start this section with a short discussion concerning preliminary preprocessing of time series; this can also be considered as a part of the SSA processing.

2.5.1 Preprocessing

There are two standard ways of preprocessing, log-transformation and centering. The log-transformation has already been discussed in Sect. 2.3.1.3. It is a very important feature of SSA that even if the main model of the series is multiplicative, SSA can work well without the use of the log-transformation. It is an essential advantage of SSA over many other methods as full multiplicativity is a very strong assumption and generally it is not met in practice. For example, the time series 'War', 'US unemployment' and 'Germany unemployment' are similar to multiplicative time series. However, the log-transformation does not provide constancy of seasonal amplitudes while the main assumption of many conventional methods is similarity of the seasonal components from year to year.

Centering of time series (that is, the subtraction of the general mean) is necessary for the application of methods of analysis of stationary time series. Usually, these methods deal with estimation of spectral characteristics of time series. This means that centering has little sense for time series with trends. From the viewpoint of SSA, centering can both increase and decrease the rank of the time series. For example, the trend of 'Fortified wine' (Sect. 2.3.1.2, Fig. 2.16) is very well described by one leading eigentriple with share 94.6% ($L = 84$), i.e., it is well approximated by an exponential series of rank 1. After centering, the trend is described by two eigentriples ET3 (11.2%) and ET4 (8.9%). The accuracy of trend extraction is worse and the extraction of trend is more complicated since the corresponding eigentriples are no longer the leading eigentriples.

Sometimes the centering of the series may be very useful. As an example, consider the time series 'S&P500', the free-float capitalization-weighted index of the prices of 500 large-cap common stocks actively traded in the United States. Its trend has complex form. However, in the timeframe of 1.5 year the trend of the centered series can be approximated by a sinusoid. The resultant trends are depicted in Fig. 2.28 together with the initial series. The first trend is extracted from the initial series by ET1–3 (red thick line), the second trend is extracted from the centered series by ET1–2 (green line with dots), $L = 170$. The former trend is more detailed, while the latter one is more stable.

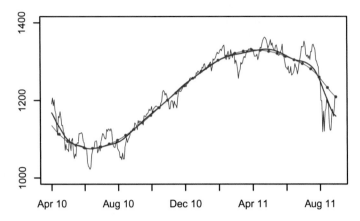

Fig. 2.28 S&P500: trends extracted from the initial series and from the centered series

2.5.2   Prior and Posterior Information in SSA

Let us consider the information about the time series that can help either to modify the SSA algorithm for more accurate estimates or to improve the analysis of the SSA results. Note that the following general rule applies: if the used prior information is wrong, the modified algorithm is worse than the original and may even yield totally wrong results.

The most frequently used prior assumption is the stationarity of the time series. In this case, Toeplitz SSA of Sect. 2.5.2.2 is used.

Another possible prior assumption is that the trend is polynomial. Especially for the linear trend, it is theoretically proven and empirically confirmed that SSA with projection can considerably improve the trend extraction and is superior to the linear regression, see Sect. 2.5.2.1 for details.

The second approach can be called posterior [27, 36] or bootstrap. On the basis of the bootstrap approach, bootstrap confidence intervals can be constructed for any characteristic estimated by SSA. The bootstrap approach includes estimation of the signal and noise parameters based on the SSA decomposition. Then simulation of a sample consisting of the estimated signal plus simulated noise allows one to construct confidence and prediction intervals. Note that the same approach is used in Monte Carlo SSA (which is actually Bootstrap SSA) for testing hypotheses, see Sect. 1.3.6, and in the SSA forecasting for constructing the bootstrap confidence intervals [21, Sect. 3.2.1.5]. The posterior approach used in [27, 36] for the detection of trend/periodic components tests the stability of the decomposition components to distinguish between signal and noise.

2.5.2.1 Centering in SSA

Consider the following extension of Basic SSA. Assume that we have selected the window length L. For $K = N - L + 1$, consider a matrix $\mathbf{A}$ of dimension $L \times K$ and rather than using the trajectory matrix $\mathbf{X}$ of the series $\mathbb{X}$ we shall use the matrix $\mathbf{X}^\star = \mathbf{X} - \mathbf{A}$. Let $\mathbf{S}^\star = \mathbf{X}^\star(\mathbf{X}^\star)^T$, and denote by λ_i and U_i $(i = 1, \ldots, d)$ the nonzero eigenvalues and the corresponding orthonormal eigenvectors of the matrix $\mathbf{S}^\star$. Setting $V_i = (\mathbf{X}^\star)^T U_i / \sqrt{\lambda_i}$ we obtain the decomposition

$$\mathbf{X} = \mathbf{A} + \sum_{i=1}^{d} \mathbf{X}_i^\star \qquad (2.24)$$

with $\mathbf{X}_i^\star = \sqrt{\lambda_i} U_i V_i^T$, instead of the standard SVD (2.2). At the grouping stage, the matrix $\mathbf{A}$ will enter one of the resultant matrices as an addend. In particular, it will produce a separate time series component after diagonal averaging.

If the matrix $\mathbf{A}$ is orthogonal to all $\mathbf{X}_i^\star$, then the matrix decomposition (2.24) yields the decomposition $\|\mathbf{X}\|_F^2 = \|\mathbf{A}\|_F^2 + \sum_{i=1}^{d} \|\mathbf{X}_i^\star\|_F^2$ of the squared norms of the corresponding matrices. Then $\|\mathbf{A}\|_F^2 / \|\mathbf{X}\|_F^2$ corresponds to the share of $\mathbf{A}$ in the decomposition.

Let us briefly consider two ways of choosing the matrix $\mathbf{A}$, both of which are thoroughly investigated in [18, Sects. 1.7.1 and 6.3].

Single centering is the row centering of the trajectory matrix. Here $\mathbf{A} = [E(\mathbf{X}) : \ldots : E(\mathbf{X})]$, where i-th component of the vector $E(\mathbf{X})$ $(i = 1, \ldots, L)$ is equal to the average of the i-th components of the lagged vectors $X_1, \ldots, X_K$. Basic SSA with single centering is expected to be superior to the standard Basic SSA if the series $\mathbb{X}$ has the form $\mathbb{X} = \mathbb{X}^{(1)} + \mathbb{X}^{(2)}$, where $\mathbb{X}^{(1)}$ is a constant series and $\mathbb{X}^{(2)}$ oscillates around zero.

For *double centering*, the SVD is applied to the matrix computed from the trajectory matrix, by subtracting from each of its elements the corresponding row and column averages and by adding the total matrix average. Basic SSA with double centering can outperform the standard Basic SSA if the series $\mathbb{X}$ can be expressed in the form $\mathbb{X} = \mathbb{X}^{(1)} + \mathbb{X}^{(2)}$, where $\mathbb{X}^{(1)}$ is a linear series (that is, $x_n^{(1)} = an + b$) and $\mathbb{X}^{(2)}$ oscillates around zero. As shown in [18, Sects. 1.7.1 and 6.3] theoretically and confirmed in [16] numerically, Basic SSA with double centering can have serious advantage over linear regression.

Centering can be considered as subtracting the projection of columns/rows onto the vector of ones. In [16], this approach is extended to incorporate projections to arbitrary subspaces. The R-code for SSA with projection is given in [21, Sect. 2.3.4].

2.5.2.2 Stationary Series and Toeplitz SSA

If the length N of the time series $\mathbb{X}$ is not sufficiently large and the series is assumed stationary, then the usual recommendation is to replace the matrix $\mathbf{S} = \mathbf{X}\mathbf{X}^{\mathrm{T}}$ by some other matrix, which is constructed under the stationarity assumption.

Note first that we can consider the *lag-covariance matrix* $\mathbf{C} = \mathbf{S}/K$ instead of $\mathbf{S}$ for obtaining the SVD of the trajectory matrix $\mathbf{X}$. Indeed, the eigenvectors of the matrices $\mathbf{S}$ and $\mathbf{C}$ are the same. Denote by $c_{ij} = c_{ij}(N)$ the elements of the lag-covariance matrix $\mathbf{C}$. If the time series is stationary, and $K \to \infty$, then $\lim c_{ij} = R_{\mathbb{X}}(|i - j|)$ as $N \to \infty$, where $R_{\mathbb{X}}(k)$ stands for the lag k term of the time series covariance function. We can therefore define a Toeplitz version of the lag-covariance matrix by putting equal values $\widetilde{c}_{ij}$ at each matrix antidiagonal $|i - j| = k$. The most natural way for defining the values $\widetilde{c}_{ij}$ and the corresponding matrix $\widetilde{\mathbf{C}}$ is to compute

$$\widetilde{c}_{ij} = \frac{1}{N - |i - j|} \sum_{m=1}^{N-|i-j|} x_m x_{m+|i-j|}, \quad 1 \leq i, j \leq L. \tag{2.25}$$

The R-code for Toeplitz SSA is provided in [21, Sect. 2.2.3].

If the original time series is stationary, the use of *Toeplitz lag-covariance matrix* $\widetilde{\mathbf{C}}$ can be more appropriate than the use of the lag-covariance matrix $\mathbf{C}$. However, Toeplitz SSA is not appropriate for nonstationary time series and if the original series has an influential nonstationary component; for such series, Basic SSA seems to work better than Toeplitz SSA. For example, if we are dealing with a pure exponential series, then it is described by a single eigentriple for any window length, while Toeplitz SSA produces L eigentriples for the window length L with harmonic-like eigenvectors. The same effect takes place for the linear series, exponential-cosine series, etc.

A number of papers devoted to SSA analysis of climatic time series (e.g. [10]) consider Toeplitz SSA as the main version of SSA and state that the difference between the Basic and Toeplitz versions of SSA is marginal. This is, however, not true if the series we analyze is non-stationary. In particular, the Toeplitz version of SSA is unsafe if the series contains a trend or oscillations with increasing or decreasing amplitudes. In Fragment 2.2.2 of [21], Toeplitz SSA is compared with Basic SSA depending on the exponential rate of series modulation (the rate 0 corresponds to stationarity). Examples of effects observed when Toeplitz SSA is applied to non-stationary time series are presented in [12]. For the study of theoretical properties of Toeplitz SSA, see, for example [23].

2.5.3 Rotations for Separability

2.5.3.1 Approach

The SVD is the key step in SSA; it provides the best matrix approximations to the trajectory matrix $\mathbf{X}$. The SVD often delivers the proper decomposition from the viewpoint of weak separability. However, if several components of the original series are mixed in such a way that their contributions are very similar, then the optimality of the SVD does not help in separating these components and we find ourselves in the situation where we have weak separability of components but are lacking their strong separability. In this situation, we need to find special rotations which would allow us to separate the components.

The examples of SSA modifications with nested decompositions are presented below; the first two items are related to rotations for improving strong separability, the last one deals with improving the weak separability. (For a detailed description of the approach that uses rotations and nested decompositions, see [21, Sect. 1.2.1].)

- *SSA-ICA.* Modification of SSA related to ICA can help if the components of the signal were mixed due to equal contributions of components (that is, when the components are weakly but not strongly separable). SSA with the SOBI-AMUSE version of ICA is described in [14]. The version of SSA with negentropy maximization is considered in Sect. 2.5.3.2, which also contains a discussion on the relation between the SSA and ICA methods.
- *SSA with derivatives.* This modification is also related to improving the strong separability. It is described in [15]. The approach changes the contribution of signal components by taking sequential differences of the series; these differences are discrete analogues of derivatives. The main reason why this can work well is the fact that the trajectory space of time series consisting of sequential differences of a signal lies in the trajectory space of this signal. Thus, the signal subspace is not corrupted but the contributions of the signal components become different, even if these contributions were equal in the straightforward versions of SSA that do not use derivatives. The R-code for SSA with derivatives is provided in [21, Sect. 2.5.4].
- *Iterated Oblique SSA.* The approach is based on oblique rotations; this version helps to solve the problem of no orthogonality (no weak separability), see [15]. The orthogonality of two vectors X and Y means that their Euclidean inner product is equal to zero, that is, $\langle X, Y \rangle = X^{\mathrm{T}}Y = 0$. Oblique SSA allows one to separate signal components, whose subseries are orthogonal with respect to the inner product $\langle X, Y \rangle_{\mathbf{A}} = X^{\mathrm{T}}\mathbf{A}Y$, where $\mathbf{A}$ is a symmetric positive definite matrix. To find an appropriate $\mathbf{A}$, iterations are used. The R-code for Iterated Oblique SSA is given in [21, Sect. 2.4.4].

2.5.3.2   SSA–ICA with Projection Pursuit

Below in this section, we shall use the approach standard in the projection pursuit method of multivariate analysis (see [28] for a review). For choosing directions, the projection pursuit uses a criterion based on the form of the distribution of the projection on a given direction. Assuming $L \leq K$, we apply the projection pursuit to the trajectory matrix with its rows considered as variables.

Let us start by considering projections on different directions for two vectors taken from subspaces corresponding to different time series components. For simplicity of illustration, we rotate the data and consider projections on the x-axis. Figure 2.29c shows projections for different rotations of two sine wave variables. The first picture in a row (the case $\alpha = 0$) corresponds to the proper rotation, the last one (with $\alpha = \pi/4$) shows the worst possible mixture. We can see that the estimated densities are totally different. To check that this result is generic, let us consider similar pictures for a sine wave and a linear function (Fig. 2.29d). The result is very similar. We thus conclude that the idea of projection pursuit may help in solving the problem of separation.

The projection pursuit method is used in ICA [29]. The aim of the ICA is finding statistically independent components $\{\eta_i;\ i = 1, ..., p\}$ from observations of their linear combinations $\{\xi_i;\ i = 1, ..., p\}$.

It is shown in [29] that in order to find independent components $\{\eta_i;\ i = 1, ..., p\}$ we need to minimize the mutual information or maximize the total negentropy over the set of all linear combinations of the initial variables. Rather than maximizing negentropies, one can consider maximization of simple functionals [29]. One of such functionals (see the first edition of the present book for details) is implemented in the R-package *fastICA* [37]; this version will be considered for Example 2.2.

In applications to blind signal separation, the cooperation between SSA and ICA has been already considered, see [41]. In this application, Basic SSA is used for

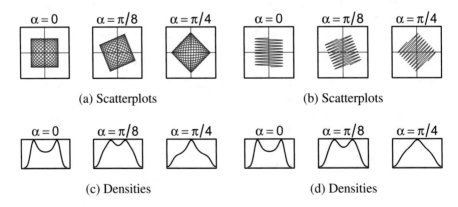

(a) Scatterplots   (b) Scatterplots

(c) Densities   (d) Densities

Fig. 2.29 Projection: two sine waves (left) and sine wave with linear series (right)

removal of noise and then the ICA is applied for extracting independent components from the mixture.

The theory of ICA is developed for random variables and is not applicable in the deterministic case. Therefore, the application of the ICA to deterministic sources can be formally considered as the projection pursuit which searches for the linear combination of the observed variables (the factor vectors in SSA) that maximizes an appropriate functional. Since the concept of statistical independence is not defined for deterministic vectors we will use the names 'ICA' and 'independent vectors' purely formally and may use quotes while referring to them. It has been established by computer simulations and confirmed by theoretical results that in the examples considered in Fig. 2.29 and some similar ones, the 'ICA' does indeed succeed in separating the time series components, even if the SVD does not provide strong separability.

The 'ICA' has the following important drawback: it does not make ordering of the found components (vectors) like the SVD does. In particular, two vectors corresponding to a sine wave can have arbitrary numbers in the decomposition by the ICA and therefore searching for them is a more difficult task than while applying the SVD. Also, the accuracy of weak separability, which the ICA provides, is worse than that for the SVD. Moreover, the numerical stability of the ICA procedures is worse than for the SVD. Therefore, in SSA, the ICA is worthwhile to consider only as a supplement to the SVD for finding proper rotations in the presence of weak separability but the lack of strong separability. By no means the ICA can be recommended as a full replacement of the SVD.

Below we suggest a scheme for building a refined grouping by the SSA–ICA procedure. This scheme could be used as a substitution of the grouping step in Basic SSA. Note that the scheme of refined grouping provided below is the same for all methods of nested decompositions for improving separability; such methods have been discussed in Sect. 2.5.3.1.

Refined Grouping by SSA–ICA
Assume we have the expansion $\mathbf{X} = \sum_{j=1}^{d} \sqrt{\lambda_j} U_j V_j^T$ at the SVD step.

1. Make a grouping $\mathbf{X} = \mathbf{X}_{I_1} + \ldots + \mathbf{X}_{I_m}$ as in Basic SSA; this corresponds to weakly separated time series components.
2. Choose a group I consisting of p indices, which is possibly composed of several interpretable components that are mixed.
3. Construct a nested decomposition by applying the 'ICA' to $\mathbf{X}_I$ and extracting p 'independent' vectors Q_i. Then $\mathbf{X}_I = \sum_{i=1}^{p} P_i Q_i^T$, where $P_i = \mathbf{X}_I Q_i$.
4. Make k subgroups from the group I by splitting $\mathbf{X}_I = \mathbf{X}_{I,1} + \ldots + \mathbf{X}_{I,k}$.

Example 2.2 Let us provide an example of application of the algorithm of SSA–ICA. Consider the example 'Fortified wines' depicted in Fig. 2.16. For the analysis, we take the first 120 points. The window length L does not provide strong separability for ET8–11 (sine waves with periods 2.4 and 3), see Fig. 2.30a depicting the w-correlation matrix, where the block of four correlated components is clearly seen. 2D-scatterplots of factor vectors are depicted in Fig. 2.30c and demonstrate the absence of

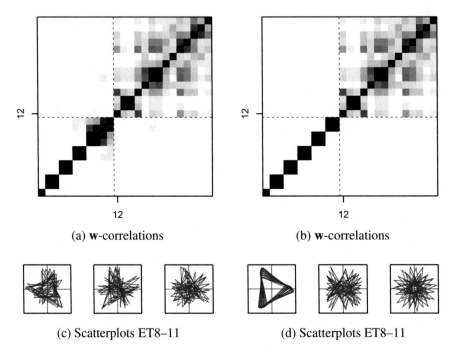

(a) **w**-correlations   (b) **w**-correlations

(c) Scatterplots ET8–11   (d) Scatterplots ET8–11

Fig. 2.30 'Fortified wines': SVD (left) and ICA for separability ET8-11 (right)

structure. Let us apply 'ICA' to the trajectory matrix reconstructed by the eigentriples 8–11. Figure 2.30b and d show that the 'ICA' makes a successful separation of the two sine waves. Let us remark that the resultant components of the 'ICA' needed an additional ordering so that the two sine waves with the same frequency obtain consecutive indices.

2.5.4 Sequential SSA

The hurdle of mixed time series components (formally, the problem of close singular values for weakly separable series components) may sometimes be overcome by the use of what was called in [18] *Sequential SSA* (alternative names for this procedure would be 'Multi-stage SSA' or 'Reiterated SSA').

The Sequential SSA with two stages can be described as follows. First, we extract several time series components by Basic SSA (or any other version of SSA) with certain window length L_1. Then we apply Basic SSA with window length L_2 to the residuals. Having extracted two sets of time series components, we can group them in different ways. For instance, if a rough trend has been extracted at the first stage and other trend components were found at the second stage, then we have to add them

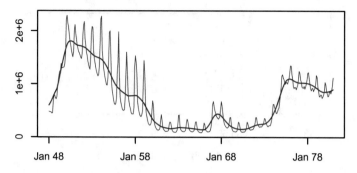

Fig. 2.31 Germany Unemployment: the initial series and its trend

together to obtain a refined accurate trend. Let us illustrate this on the following example.

Example 2.3 *'Germany Unemployment' series: extraction of harmonics*
The 'Germany unemployment' time series (West Germany, monthly, from April 1950 to December 1980, [42]) serves as an example of complex trends and amplitude-modulated periodicities. The time series is depicted in Fig. 2.31.

Selecting large L would mix up the trend and periodic components of the series. For small L the periodic components are not separable from each other. Hence Basic SSA fails to extract (amplitude-modulated) harmonic components of the series.

The Sequential SSA with two stages is a better method in this case. If we apply Basic SSA with $L = 12$ to the initial series, then the first eigentriple will describe the trend (see Fig. 2.31) which is extracted rather well: the trend component does not include high frequencies, while the residual component practically does not contain low ones (see Fig. 2.32a for the residual series).

The second stage of Sequential SSA is applied to the residual series with $L = 180$. Since the time series is amplitude modulated, the main periodogram frequencies (annual $\omega = 1/12$, half-annual $\omega = 1/6$ and 4-months $\omega = 1/4$) are somewhat spread out, and therefore each (amplitude-modulated) harmonic can be described by several (more than 2) eigentriples.

The periodogram analysis of the obtained singular vectors shows that the leading 14 eigentriples with share 91.4% can be related to the following three periodicities: the eigentriples 1, 2, 5–8, 13, 14 describe the annual amplitude-modulated harmonic (Fig. 2.32b), the eigentriples 3, 4, 11–12 are related to half-year periodicity, and the eigentriples 9, 10 describe the 4-months harmonic. □

The same technique can be applied to the 'Births' series if we want to obtain better results than those described in Sect. 2.2.2.2. (See Sect. 2.4.3 for a discussion concerning the large window length problem in this example.)

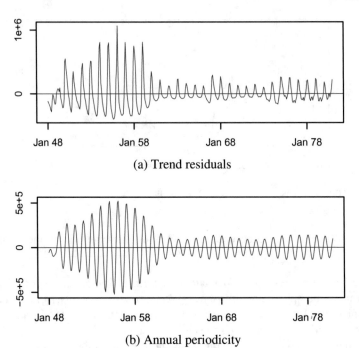

(a) Trend residuals

(b) Annual periodicity

Fig. 2.32 'Germany unemployment': oscillations

2.5.5 SSA and Outliers

The problem of robustness to outliers is essential for any technique of data analysis. Let us consider how this problem can be solved in SSA. Recall that for signal extraction, Basic SSA can be expressed by (2.22) of Sect. 2.4.4, where the main two operators are the projections Π_r and $\Pi_{\mathcal{H}}$. In Basic SSA, both projections are performed in the Frobenius norm, which can be called the L_2-norm.

There exist two robust modifications of SSA, in which the projections are performed in other norms.

Weighted projections. The first approach allows the use of a weighted norm, where different weights are given to the different time series points; the points which are suspected to be outliers have smaller weights. In this approach, the projections are performed in the weighted Frobenius norm. The weights are chosen by an iterative procedure like that used in the LOWESS nonparametric smoothing [9] or in the iteratively reweighted least-squares method (IRLS) [26], where the weights are iteratively chosen depending on the residuals.

In this approach, the algorithm of SSA with weights should be implemented. SSA with special weights, where the ordinary SVD is changed to the oblique SSA, can be used with approximately the same computational cost as Basic SSA [56]. However, this is not the case if time series points have arbitrary weights. The SVD with arbitrary weights has no closed-form solution and therefore needs an iterative numerical solution (Sect. 3.8.4 contains discussions on how to deal with weighted SVD and on Cadzow iterations whose weighted versions are similar to this robust algorithm except for recalculation of weights). Thus, the algorithm of SSA with weighted projections, which helps to remove outliers, contains two loops: the inner loop (to calculate the weighted SVD) and the outer loop (to recalculate weights basing on the residuals); as a result, such algorithm is very time-consuming. This approach is described in [48], where the authors consider the weighted projection to the set of low-rank matrices. Note that the weighted projection should be used also for the projection to the set of Hankel matrices [56, Proposition 2].

L_1 projections. The second approach is also frequently used in approximation problems. To improve the robustness, the projections are constructed in the L_1-norm. The idea to use the L_1-norm matrix approximation instead of the L_2-norm one (that is, instead of the ordinary SVD if we talk about SSA) is very popular in data analysis. As L_1-SVD has no closed-form solution, time-consuming iterative algorithms should be applied, see the brief discussion in [13, Sect. 3.12]. The L_1-projection on the set of Hankel matrices is performed by the substitution of the diagonal averaging to taking medians instead of averages. The algorithm based on the use of L_1-norm is considered in [32, 43]. However, the problem of its implementation with reasonable computational cost is not yet resolved.

2.5.6 Replacing the SVD with Other Procedures

Some variations to the standard SVD procedure have been already mentioned in Sects. 2.5.3 and 1.4. These variations include rotations within the eigenspaces, Independent Component Analysis (ICA) and Partial SVD where only few leading eigenvectors of the matrix $\mathbf{S} = \mathbf{X}\mathbf{X}^T$ are computed.

There are four main reasons why it may be worthwhile to replace the SVD operation in Basic SSA with some other operation. These reasons are: (a) simplicity, (b) improved performance, (c) different optimality principles for the decomposition, and (d) improved robustness.

(a) Simplicity. This is important in the problems where the dimension of the trajectory matrix is very large. In these cases, the SVD may be too costly to perform. The most obvious substitution of the SVD is by Partial SVD mentioned above. Another useful substitution of the SVD is suggested in [38] for problems of the form 'signal plus noise'. In this paper, a polynomial approximation $P(\mathbf{X})$ to $\mathbf{X}^{(1)}$ is developed in the model $\mathbb{X} = \mathbb{X}^{(1)} + \mathbb{X}^{(2)}$, where we fix the cut-off point in this spectrum of $\mathbf{X}$ which would distinguish $\mathbf{X}^{(1)}$ from $\mathbf{X}^{(2)}$. Many numerical approximations for the solution of the full or partial SVD problem are also available, see [11] and Sect. 1.4. In cases where the dimension of the matrix $\mathbf{S}$ is exceptionally large, one can use the approximations for the leading eigenvectors used in internet search engines, see e.g. [35].

(b) Improved performance. In some cases (usually when a parametric form of the signal is given), one can slightly correct the SVD (both, eigenvalues and eigenvectors) using the recommendations of SSA perturbation theory, see the end of Sect. 2.3.3. As a simple example, in the problems of separating signal from noise, some parts of noise are often found in the SVD components mostly related to the signal, see Fig. 2.22a and b. As a result, it may be worthwhile to make small adjustments to the eigenvalues and eigenvectors to diminish this effect. The simplest version of Basic SSA with constant adjustment in all eigenvalues was suggested in [50] and is sometimes called the minimum-variance SSA.

(c) Different optimality principles. Here the basis for the decomposition of the series is chosen using some principle which is different from the SVD optimality. For example, in ICA discussed in Sect. 2.5.3, the independence of components (rather than the precision of approximation) is considered as the main optimality criteria.

(d) Improved robustness. Different robust versions of the SVD can be considered if outliers are expected in the observed time series, see Sect. 2.5.5.

2.5.7 Complex SSA

Complex SSA is Basic SSA applied to complex-valued time series. The only difference in the algorithms is the use of the conjugate transpose rather than the usual transpose. The algorithm of Complex SSA was explicitly formulated in [34] although SSA has been applied to complex time series much earlier, without considering the difference between real-valued and complex-valued data (see, e.g., [49]). The R-code for the call of Complex SSA can be found in [21, Sect. 4.1.4]. Note that recently a more effective implementation was added to RSSA with the use of the svd method from the R-package PRIMME [55]; one can set the parameter svd.method = "primme" in the call of ssa.

Despite Complex SSA can be considered as a version of SSA for the analysis of two real time series [17, 20], it is a more appropriate method for analyzing one-dimensional complex time series. Most applications of Complex SSA are related to the so-called F-xy eigenimage filtering [47]. This name is related to the analysis of digital images in geophysics; first, the discrete Fourier transform (DFT) is applied to each row of the image matrix, then complex-valued series are constructed from the results of the DFTs for each frequency, and finally, the constructed series are analyzed by Complex SSA. Note that the authors of the papers devoted to F-xy eigenimage noise suppression usually omit the word 'Complex' in Complex SSA. The specificity of the studied geophysical images is that they are noisy and contain lines (traces); with the help of the DFT, these lines are transformed into a sum of complex exponentials, which have rank 1 in Complex SSA. Therefore, the lines can be separated from noise very well.

2.6 Multidimensional and Multivariate Extensions of SSA

Most often, SSA is applied to time series, or 1D (one-dimensional) sequences. However, it is easy to extend SSA for analysing multidimensional objects such as collections of time series and digital images, which are not necessarily rectangular (for a detailed description, see [21, Sect. 1.2.2]). All SSA algorithms have the following common sequence of operations: (1) transformation of the initial digital object into a matrix (called trajectory matrix) by an embedding operator, (2) decomposition of the trajectory matrix into elementary matrices, (3) grouping the elementary matrices, and (4) obtaining the decomposition of the initial object, which is interpretable, if the grouping is made properly.

Multivariate/multidimensional extensions differ by the embedding step only in the general SSA scheme. In Basic SSA, the embedding step corresponds to transformation of the original time series into a Hankel matrix. Denoting this transformation of the input object by the embedding operator $\mathcal{T}$, one can say that multidimensional extensions of SSA differ by the form of $\mathcal{T}$. For all extensions considered below, the embedding operator $\mathcal{T}$ transforms the initial object into a matrix, although there are modifications that consider embedding in the form of tensors, see a brief review in [13, Sect. 2.10.3]. Let us list different versions of the embedding step:

- SSA for time series (1D-SSA): $\mathcal{T}(\mathbb{X})$ is a Hankel matrix;
- MSSA for collections of time series: $\mathcal{T}(\mathbb{X})$ is a stacked Hankel matrix;
- 2D-SSA for digital images: $\mathcal{T}(\mathbb{X})$ is a Hankel-block-Hankel matrix;
- Shaped SSA for any shaped objects: $\mathcal{T}(\mathbb{X})$ is a quasi-Hankel matrix.

Shaped nD-SSA ($n \geq 1$) can be considered as a general n-dimensional extension of SSA for shaped objects. The original 1D-SSA and all other extensions of SSA are particular cases of Shaped nD-SSA. This observation is helpful from the theoretical viewpoint, since it allows us to consider the SSA theory in a general form. This is also helpful for effective implementation of the SSA algorithms, since it is enough to have an effective implementation of Shaped nD-SSA.

This general approach is used in [20, 21, 44]. Below we consider particular cases to explain how the SSA algorithms are changing in the multidimensional modifications. We do not have to write down the whole algorithm, since the algorithm is the same up to a description of the embedding operator; that is, up to a description of the way of constructing the trajectory matrix. Indeed, we have mentioned in Sect. 2.1.2.4 that the step of diagonal averaging corresponds to the projection of matrices to the set of Hankel matrices in the Frobenius norm. In a similar manner, if the embedding operator provides a one-to-one correspondence between the set of input objects and a subspace of matrices of special kind, the projection to this subspace is used to obtain the decomposition of the initial object from the matrix decomposition in the multidimensional case. We can also refer to the condensed form (2.22) of Basic SSA for signal extraction and note that (2.22) also holds for all multivariate and multidimensional extensions if we use suitable $\mathcal{T}$ and $\Pi_{\mathcal{H}}$, which is no longer a projection to the set of Hankel matrices but rather to the set of matrices $\{\mathcal{T}(\mathbb{Z}), \forall \mathbb{Z}\}$.

2.6.1 MSSA

In MSSA (recall that MSSA is an abbreviation for Multivariate/Multichannel SSA), the trajectory matrix is constructed from the stacked trajectory matrices of time series from the considered collection. The stacking can be either vertical or horizontal. We consider the horizontal stacking as the basic version (see also [20, 21]):

$$\mathcal{T}_{MSSA}(\mathbb{X}^{(1)}, \ldots, \mathbb{X}^{(s)}) = [\mathbf{X}^{(1)} : \ldots : \mathbf{X}^{(s)}],$$

where s is the number of 1D series, $\mathbf{X}^{(i)}$ is the trajectory (and therefore Hankel) matrix of the i-th time series $\mathbb{X}^{(i)}$; note that time series lengths can differ. Therefore, the left singular vectors belonging to the column space correspond to subseries of length L, while the right singular vectors belonging to the row space consist of stacked subseries of different time series from the collection. In contrast to 1D-SSA, left and right singular vectors have different structure.

In the case of vertical stacking (for equal series lengths),

$$\mathcal{T}_{MSSA}(\mathbb{X}^{(1)}, \ldots, \mathbb{X}^{(s)}) = \begin{pmatrix} \mathbf{X}^{(1)} \\ \ldots \\ \mathbf{X}^{(s)} \end{pmatrix}. \tag{2.26}$$

The vertical stacking is not fundamentally different from the horizontal stacking; the left and right singular vectors interchange (after interchanging L and $K = N - L + 1$) and therefore, the way of stacking influences no more than terminology, the choice of parameters and possibly computational costs. It is important to bear in mind that there are many different notations in literature on MSSA, and some of these notations are somewhat controversial, see discussions in [21, Sects 4.2,4.3], [13] and Sect. 3.10.1.

With the horizontal stacking, we fix the number of rows; that is, the dimension of the column space of the trajectory matrix. In this case, the increase of the time series lengths leads to the increase in the number of columns of the trajectory matrix. Recall that for the horizontal stacking, the lengths of different time series can differ and this has no influence on the column dimension.

For a simple illustration, consider the horizontal and vertical stackings for two time series, $\mathbb{X}^{(1)} = (11, 12, 13, 14, 15)$ and $\mathbb{X}^{(2)} = (21, 22, 23, 24)$. For the horizontal stacking with $L = 2$ we have

$$\begin{pmatrix} 11 & 12 & 13 & 14 & | & 21 & 22 & 23 \\ 12 & 13 & 14 & 15 & | & 22 & 23 & 24 \end{pmatrix}.$$

The vertical stacking is possible only if the number of columns in the individual trajectory matrices is the same; in our example, therefore, we have to use different window lengths for different time series; set $L_1 = 2$ and $L_2 = 1$:

$$\begin{pmatrix} 11 \ 12 \ 13 \ 14 \\ 12 \ 13 \ 14 \ 15 \\ \overline{21 \ 22 \ 23 \ 24} \end{pmatrix}.$$

Let us explain the difference between the vertical staking in the form (2.26) and in the form used in [54]. Let $\mathbb{X}^{(j)} = (j1, j2, j3, j4, j5)$, where $j = 1, \ldots, 3$ are numbers of some geographic points, the second digit corresponds to time. Then the rows of the trajectory matrix $\mathbf{X}$ for MSSA with vertical stacking are ordered by series numbers, whereas the rows of the trajectory matrix $\mathbf{X}_{\mathrm{WN}}$ used in [54] are ordered by time:

$$\mathbf{X} = \begin{pmatrix} 11 \ 12 \ 13 \ 14 \\ 12 \ 13 \ 14 \ 15 \\ 21 \ 22 \ 23 \ 24 \\ 22 \ 23 \ 24 \ 25 \\ 31 \ 32 \ 33 \ 34 \\ 32 \ 33 \ 34 \ 35 \end{pmatrix}, \quad \mathbf{X}_{\mathrm{WN}} = \begin{pmatrix} 11 \ 12 \ 13 \ 14 \\ 21 \ 22 \ 23 \ 24 \\ 31 \ 32 \ 33 \ 34 \\ 12 \ 13 \ 14 \ 15 \\ 22 \ 23 \ 24 \ 25 \\ 32 \ 33 \ 34 \ 35 \end{pmatrix}. \tag{2.27}$$

However, the SVD decompositions of the two trajectory matrices in (2.27) are in fact the same and differ only in the order of appearance of the vector components. The left singular vectors of size $6 = 2 \times 3$ can be split into two parts shifted in time. Each part consists of three geographic points. In [54], the left singular vectors of length Ls are called EEOFs (extended empirical orthogonal functions). In (2.27), each EEOF is divided into two parts corresponding to two time lags (the number of lags is equal to $L = 2$) and then each part is depicted as a surface.

The R-code for MSSA is contained in [21, Sect. 4.2.6].

2.6.2 2D-SSA

2D-SSA is a technique of decomposition of digital images (see overviews in [20] and [21, Chap. 5]).

For a digital image given by a 2D array

$$\mathbb{X} = \mathbb{X}_{N_1, N_2} = (x_{ij})_{i,j=1}^{N_1, N_2} = \begin{pmatrix} x_{1,1} \ \cdots \ x_{1,N_2} \\ \cdots \ \cdots\cdots\cdots \\ x_{N_1,1} \ \cdots \ x_{N_1,N_2} \end{pmatrix}$$

of size $N_1 \times N_2$, a 2D window size $L_1 \times L_2$ should be chosen.

The trajectory matrix consists of vectorized moving 2D windows. In 2D-SSA, the columns of the trajectory matrix are the vectorized windows of size $L_1 \times L_2$, while the rows are the vectorized windows of size $K_1 \times K_2$, where $K_i = N_i - L_i + 1$. Thus, the trajectory matrix has size $L_1 L_2 \times K_1 K_2$. Such trajectory matrix can be written in the form of a Hankel-block-Hankel matrix:

$$
\mathbf{X} = \mathcal{T}_{\text{2D-SSA}}(\mathbb{X}) =
\begin{pmatrix}
\mathbf{H}_1 & \mathbf{H}_2 & \mathbf{H}_3 \cdots & \mathbf{H}_{K_2} \\
\mathbf{H}_2 & \mathbf{H}_3 & \mathbf{H}_4 \cdots & \mathbf{H}_{K_2+1} \\
\mathbf{H}_3 & \mathbf{H}_4 & \ddots & \vdots \\
\vdots & \vdots & \ddots & \vdots \\
\mathbf{H}_{L_2} & \mathbf{H}_{L_2+1} & \cdots \cdots & \mathbf{H}_{N_2}
\end{pmatrix} ;
\tag{2.28}
$$

here each $\mathbf{H}_j$ is the $L_1 \times K_1$ trajectory (Hankel) matrix constructed from $\mathbb{X}_{:,j}$ (the j-th column of the 2D array $\mathbb{X}$). Here we have moved the window from top to bottom and then from left to right.

For example, let us take an image

$$
\mathbb{X} =
\begin{pmatrix}
1 & 4 & 7 \\
2 & 5 & 8 \\
3 & 6 & 9
\end{pmatrix}
$$

and the window of size 2×2. Then we have a set of four windows $\begin{pmatrix} 1 & 4 \\ 2 & 5 \end{pmatrix}$, $\begin{pmatrix} 2 & 5 \\ 3 & 6 \end{pmatrix}$, $\begin{pmatrix} 4 & 7 \\ 5 & 8 \end{pmatrix}$, $\begin{pmatrix} 5 & 8 \\ 6 & 9 \end{pmatrix}$ (and several additional windows in the circular case, which is described in [44]). For the ordinary case, the trajectory matrix will have the form

$$
\mathbf{X} =
\left(
\begin{array}{cc|cc}
1 & 2 & 4 & 5 \\
2 & 3 & 5 & 6 \\
\hline
4 & 5 & 7 & 8 \\
5 & 6 & 8 & 9
\end{array}
\right).
$$

One can see that the 2D trajectory matrix consists of the trajectory matrices from each matrix's column.

Let $\mathbb{X}$ be a digital image, which is modelled as a decomposition $\mathbb{X} = \mathbb{T} + \mathbb{P} + \mathbb{N}$ into a pattern, regular oscillations (e.g. a texture) and noise. The problem of estimating the decomposition components is very similar to the 1D case of time series. Figure 2.33 demonstrates the extraction of a pattern of a corrupted image; the pattern is obtained after the removal of regular oscillations.

Note that the term 'digital image' is just a common name for a 2D data set; for example, one of the two dimensions may be temporal. The decomposition problem for multidimensional data is also important in higher dimensions; these dimensions may again be of different nature. For example, one can consider both 3D spatial data and 2D data with the third temporal dimension as 3D data; and so on. We will call data with n dimensions nD data. The R-codes for application of 2D- and nD-SSA to digital images can be found in [21, Sects. 5.1.4 and 5.4.4].

MSSA for s equal-length time series of length N is a particular case of 2D-SSA with $N_1 = s$, $N_2 = N$, $L_1 = 1$, $L_2 = L$. In MSSA, the trajectory matrix has size $L \times sK$, where $K = N - L + 1$.

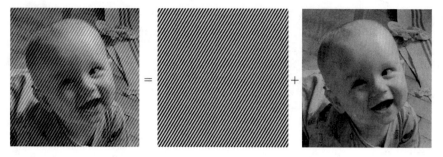

Fig. 2.33 Maya: Decomposition

2.6.3 Shaped SSA

Shaped SSA, developed in [20, 44], is the universal version of SSA, which is applicable to arbitrary shapes and dimensions of the objects with the moving window of arbitrary shape. For example, both the digital image and the moving window can be circular-shaped; that is, the image and the window in Shaped SSA do not have to be rectangular-shaped as in 2D-SSA. Shaped SSA allows one to consider different versions of the SSA algorithms in a unified manner; these versions include SSA for objects with missing data, 1D-SSA, MSSA, 2D-SSA as well as their circular variants. There is one limitation in Shaped SSA: the points within the window and the window locations within the object should be linearly ordered. Being linearly ordered, the shaped windows are transformed into column vectors and these vectors are stacked into the trajectory matrix according to the ordered locations of the shaped windows.

Many objects can be considered as linearly ordered in a natural way. The standard technique is to consider the object as a subset of a multidimensional box of the same dimension as the object box, with natural ordering. For example, a piece of a sphere (after its projection to the plane) can be circumscribed by a rectangle [46]; however, there is no continuous planar projection of the whole sphere. On the other hand, a ball or its piece is an appropriate object for the analysis by Shaped SSA (more precisely, by Shaped 3D SSA), see [45].

As an example of different linear orderings, consider the collection of $s = 3$ time series of length $N = 5$ in the form of the matrix

$$\mathbb{X} = \begin{cases} (11, \ 12, \ 13, \ 14, \ 15) \\ (21, \ 22, \ 23, \ 24, \ 25) \\ (31, \ 32, \ 33, \ 34, \ 35) \end{cases}$$

and windows of size 1×4. Columns of the trajectory matrix consist of transposed windows. By moving the window first from left to right and then from top to bottom, we obtain the transposed trajectory matrix $\mathbf{X}$ given in (2.27), which was constructed by vertical stacking of trajectory matrices of individual time series. By moving the window first from top to bottom and then from the left to the right, we obtain the transposed trajectory matrix $\mathbf{X}_{WN}$ given in (2.27). If we move the window of size 3×2 (here $3 = (s - 1) + 1$, $2 = (N - 4) + 1$), we obtain the trajectory matrix $\mathbf{X}$ in the original, non-transposed form. As we have mentioned above, different linear orderings do not change the SSA decompositions, up to the ordering of elements.

The R-code for application of Shaped SSA to objects of different shapes is given in [21, Sect. 5.2.4].

References

1. Alexandrov T (2009) A method of trend extraction using singular spectrum analysis. RevStat 7(1):1–22
2. Alonso F, Salgado D (2008) Analysis of the structure of vibration signals for tool wear detection. Mech Syst Signal Process 22(3):735–748
3. Alonso F, Castillo J, Pintado P (2005) Application of singular spectrum analysis to the smoothing of raw kinematic signals. J Biomech 38(5):1085–1092
4. Andrews D, Herzberg A (1985) Data. A collection of problems from many fields for the student and research worker. Springer, New York
5. Badeau R, Richard G, David B (2008) Performance of ESPRIT for estimating mixtures of complex exponentials modulated by polynomials. IEEE Trans on Signal Process 56(2):492–504
6. Bilancia M, Campobasso F (2010) Airborne particulate matter and adverse health events: Robust estimation of timescale effects. In: Bock HH et al (eds) Classification as a tool for research, studies in classification, data analysis, and knowledge organization. Springer, Berlin, pp 481–489
7. Brillinger D (1975) Time series. Data analysis and theory. Holt, Rinehart and Winston Inc, New York
8. Clemens J (1994) Whole earth telescope observation of the white dwarf star PG1159-035. In: Weigend A, Gershenfeld N (eds) Time series prediction: forecasting the future and understanding the past. Addison-Wesley, Reading
9. Cleveland WS (1979) Robust locally weighted regression and smoothing scatterplots. J Amer Stat Ass 74(368):829–836
10. Ghil M, Allen RM, Dettinger MD, Ide K, Kondrashov D, Mann ME, Robertson A, Saunders A, Tian Y, Varadi F, Yiou P (2002) Advanced spectral methods for climatic time series. Rev Geophys 40(1):1–41
11. Golub GH, Van Loan CF (1996) Matrix computations, 3rd edn. Johns Hopkins University Press, Baltimore
12. Golyandina N (2010) On the choice of parameters in singular spectrum analysis and related subspace-based methods. Stat Interface 3(3):259–279
13. Golyandina N (2020) Particularities and commonalities of singular spectrum analysis as a method of time series analysis and signal processing. WIREs Comput Stat 12(4):e1487
14. Golyandina N, Lomtev M (2016) Improvement of separability of time series in singular spectrum analysis using the method of independent component analysis. Vestnik St Petersburg University Mathematics 49(1):9–17
15. Golyandina N, Shlemov A (2015) Variations of singular spectrum analysis for separability improvement: non-orthogonal decompositions of time series. Stat Interface 8(3):277–294
16. Golyandina N, Shlemov A (2017) Semi-nonparametric singular spectrum analysis with projection. Stat Interface 10(1):47–57
17. Golyandina N, Stepanov D (2005) SSA-based approaches to analysis and forecast of multidimensional time series. In: Proceedings of the 5th St.Petersburg workshop on simulation, June 26-July 2, 2005, St. Petersburg State University, St. Petersburg, pp 293–298

18. Golyandina N, Nekrutkin V, Zhigljavsky A (2001) Analysis of time series structure: SSA and related techniques. Chapman&Hall/CRC, Boca Raton
19. Golyandina N, Pepelyshev A, Steland A (2012) New approaches to nonparametric density estimation and selection of smoothing parameters. Comput Stat Data Anal 56(7):2206–2218
20. Golyandina N, Korobeynikov A, Shlemov A, Usevich K (2015) Multivariate and 2D extensions of singular spectrum analysis with the Rssa package. J Stat Softw 67(2):1–78
21. Golyandina N, Korobeynikov A, Zhigljavsky A (2018) Singular spectrum analysis with R. Springer, Berlin
22. Golyandina N, Korobeynikov A, Zhigljavsky A (2018) Site-companion to the book 'Singular spectrum analysis with R'. https://ssa-with-r-book.github.io/
23. Harris T, Yan H (2010) Filtering and frequency interpretations of singular spectrum analysis. Physica D 239:1958–1967
24. Hassani H, Xu Z, Zhigljavsky A (2011) Singular spectrum analysis based on the perturbation theory. Nonlinear Anal: Real World Appl 12(5):2752–2766
25. Hipel K, McLeod A (1994) Time series modelling of water resources and environmental systems. Elsevier Science, Amsterdam
26. Holland PW, Welsch RE (1977) Robust regression using iteratively reweighted least-squares. Commun Stat - Theory Methods 6(9):813–827
27. Holmström L, Launonen I (2013) Posterior singular spectrum analysis. Stat Anal Data Min 6(5):387–402
28. Huber PJ (1985) Projection pursuit. Ann Stat 13(2):435–475
29. Hyvärinen A, Oja E (2000) Independent component analysis: algorithms and applications. Neural Netw 13(4–5):411–430
30. Ivanova E, Nekrutkin V (2019) Two asymptotic approaches for the exponential signal and harmonic noise in singular spectrum analysis. Stat Interface 12(1):49–59
31. Janowitz M, Schweizer B (1989) Ordinal and percentile clustering. Math Soc Sci 18:135–186
32. Kalantari M, Yarmohammadi M, Hassani H (2016) Singular spectrum analysis based on L1-norm. Fluctuat Noise Lett 15(01):1650,009
33. Kendall M, Stuart A (1976) Design and analysis, and time series. The advanced theory of statistics, vol 3, 3rd edn. Charles Griffin, London
34. Keppenne C, Lall U (1996) Complex singular spectrum analysis and multivariate adaptive regression splines applied to forecasting the southern oscillation. In: Exp Long-Lead Forcst Bull
35. Langville AN, Meyer CD (2005) A survey of eigenvector methods for web information retrieval. SIAM Rev 47:135–161
36. Launonen I, Holmström L (2017) Multivariate posterior singular spectrum analysis. Stat Methods & Appl 26(3):361–382
37. Marchini JL, Heaton C, Ripley BD (2019) fastICA: FastICA algorithms to perform ICA and projection pursuit. http://CRAN.R-project.org/package=fastICA, R package version 1.2-2
38. Moskvina V, Schmidt KM (2002) Approximate projectors in singular spectrum analysis. SIAM J Matrix Anal Appl 24:932–942
39. Nekrutkin V (2010) Perturbation expansions of signal subspaces for long signals. Stat Interface 3:297–319
40. Nekrutkin V, Vasilinetc I (2017) Asymptotic extraction of common signal subspaces from perturbed signals. Stat Interface 10(1):27–32
41. Pietilä A, El-Segaier M, Vigário R, Pesonen E (2006) Blind source separation of cardiac murmurs from heart recordings. In: Rosca J et al (eds) Independent component analysis and blind signal separation, vol 3889. Lecture notes in computer science. Springer, Berlin, pp 470–477
42. Rao TS, Gabr M (1984) An introduction to bispectral analysis and bilinear time series models. Springer, Berlin
43. Rodrigues P, Lourenço V, Mahmoudvand R (2018) A robust approach to singular spectrum analysis. Quality Reliabil Eng Int 34(7):1437–1447

44. Shlemov A, Golyandina N (2014) Shaped extensions of singular spectrum analysis. In: 21st International symposium on mathematical theory of networks and systems, July 7–11, 2014. Groningen, The Netherlands, pp 1813–1820

45. Shlemov A, Golyandina N, Holloway D, Spirov A (2015) Shaped 3D singular spectrum analysis for quantifying gene expression, with application to the early *Drosophila* embryo. BioMed Res Int 2015(Article ID 986436):1–18

46. Shlemov A, Golyandina N, Holloway D, Spirov A (2015) Shaped singular spectrum analysis for quantifying gene expression, with application to the early *Drosophila* embryo. BioMed Res Int 2015(Article ID 689745)

47. Trickett S (2003) F-xy eigenimage noise suppression. Geophysics 68(2):751–759

48. Trickett S, Burroughs L, Milton A (2012) Robust rank-reduction filtering for erratic noise. Technical Report, https://library.seg.org/doi/pdf/10.1190/segam2012-0129.1

49. Tufts DW, Kumaresan R, Kirsteins I (1982) Data adaptive signal estimation by singular value decomposition of a data matrix. Proc IEEE 70(6):684–685

50. Van Huffel S (1993) Enhanced resolution based on minimum variance estimation and exponential data modeling. Signal Process 33:333–355

51. Vautard R, Yiou P, Ghil M (1992) Singular-spectrum analysis: a toolkit for short, noisy chaotic signals. Physica D 58:95–126

52. Vlassieva E, Golyandina N (2009) First-order SSA-errors for long time series: model examples of simple noisy signals. In: Proceedings of the 6th St.Petersburg workshop on simulation, vol. 1, June 28-July 4, 2009, St. Petersburg, St.-Petersburg State University, pp 314–319

53. Wax M, Kailath T (1985) Detection of signals by information theoretic criteria. IEEE Trans Acoust 33:387–392

54. Weare BC, Nasstrom JS (1982) Examples of extended empirical orthogonal function analyses. Mon Weather Rev 110(6):481–485

55. Wu L, Romero E, Stathopoulos A (2017) PRIMME_SVDS: a high-performance preconditioned SVD solver for accurate large-scale computations. SIAM J Sci Comput 39(5):S248–S271

56. Zvonarev N, Golyandina N (2017) Iterative algorithms for weighted and unweighted finite-rank time-series approximations. Stat Interface 10(1):5–18

57. Zvonarev N, Golyandina N (2018) Image space projection for low-rank signal estimation: modified Gauss-Newton method. arXiv:1803.01419

Chapter 3
SSA for Forecasting, Interpolation, Filtering and Estimation

3.1 SSA Forecasting Algorithms

3.1.1 Main Ideas and Notation

A reasonable forecast of a time series can be performed only if the series has a structure and there are tools to identify and use this structure. Also, we should assume that the structure of the time series is preserved for the future time period over which we are going to forecast (continue) the series. The last assumption cannot be validated using the data to be forecasted. Moreover, the structure of the series can rarely be identified uniquely. Therefore, the situation of different and even contradictory forecasts is not impossible. Hence it is important not only to understand and express the structure but also to assess its stability.

A forecast can be made only if a model is built. The model should be either derived from the data or at least checked against the data. In SSA forecasting, these models can be described through the linear recurrence relations (LRRs). The class of series governed by LRRs is rather wide and important for practical applications. This class contains the series that are linear combinations of products of exponential, polynomial and harmonic series.

Assume that $\mathbb{X}_N = \mathbb{X}_N^{(1)} + \mathbb{X}_N^{(2)}$, where the series $\mathbb{X}_N^{(1)}$ satisfies an LRR of relatively small order and we are interested in forecasting of $\mathbb{X}_N^{(1)}$. For example, $\mathbb{X}_N^{(1)}$ can be signal, trend or seasonality. The idea of recurrent forecasting is to estimate the underlying LRR and then to perform forecasting by applying the estimated LRR to the last points of the SSA approximation of the series $\mathbb{X}_N^{(1)}$. The main assumption allowing SSA forecasting is that for a certain window length L the series components $\mathbb{X}_N^{(1)}$ and $\mathbb{X}_N^{(2)}$ are approximately strongly separable. In this case, we can reconstruct the series $\mathbb{X}_N^{(1)}$ with the help of a selected set of the eigentriples and obtain approximations to both the series $\mathbb{X}_N^{(1)}$, its trajectory space and the true LRR.

N. Golyandina and A. Zhigljavsky, *Singular Spectrum Analysis for Time Series*, SpringerBriefs in Statistics, https://doi.org/10.1007/978-3-662-62436-4_3

Let $\mathbb{X}_N = \mathbb{X}_N^{(1)} + \mathbb{X}_N^{(2)}$ and we intend to forecast $\mathbb{X}_N^{(1)}$. If $\mathbb{X}_N^{(1)}$ is a time series of finite rank $r < L$, then it generates an L-trajectory subspace of dimension r. This subspace reflects the structure of $\mathbb{X}_N^{(1)}$ and hence it can be taken as a base for forecasting.

Let us formally describe the forecasting algorithms in a chosen subspace. As we assume that estimates of this subspace are constructed by SSA, we shall refer to the algorithms as the algorithms of SSA forecasting.

Forecasting within a subspace means a continuation of the L-lagged vectors of the time series in such a way that they lie in or very close to the chosen subspace of R^L. We consider the following three forecasting algorithms: recurrent, vector and simultaneous.

Inputs in the forecasting algorithms:
  (a) Time series $\mathbb{X}_N = (x_1, \ldots, x_N)$, $N > 2$.
  (b) Window length L, $1 < L < N$.
  (c) Linear space $\mathcal{L}_r \subset \mathsf{R}^L$ of dimension $r < L$. We assume that $\mathbf{e}_L \notin \mathcal{L}_r$, where $\mathbf{e}_L = (0, 0, \ldots, 0, 1)^\mathsf{T} \in \mathsf{R}^L$; in other terms, $\mathcal{L}_r$ is not a 'vertical' space.
  (d) Number M of points to forecast for.

Notation:
  (a) $\mathbf{X} = [X_1 : \ldots : X_K]$ (with $K = N - L + 1$) is the trajectory matrix of $\mathbb{X}_N$.
  (b) $P_1, \ldots, P_r$ is an orthonormal basis in $\mathcal{L}_r$.
  (c) $\widehat{\mathbf{X}} \stackrel{\text{def}}{=} [\widehat{X}_1 : \ldots : \widehat{X}_K] = \sum_{i=1}^{r} P_i P_i^\mathsf{T} \mathbf{X}$. The vector $\widehat{X}_i$ is the orthogonal projection of X_i onto the space $\mathcal{L}_r$.
  (d) $\widetilde{\mathbf{X}} = \boldsymbol{\Pi}_{\mathcal{H}} \widehat{\mathbf{X}} = [\widetilde{X}_1 : \ldots : \widetilde{X}_K]$ is the result of hankelization of the matrix $\widehat{\mathbf{X}}$. The matrix $\widetilde{\mathbf{X}}$ is the trajectory matrix of some time series $\widetilde{\mathbb{X}}_N = (\widetilde{x}_1, \ldots, \widetilde{x}_N)$.
  (e) For any vector $Y \in \mathsf{R}^L$, we denote by $\overline{Y} \in \mathsf{R}^{L-1}$ the vector consisting of the last $L - 1$ components of the vector Y and by $\underline{Y} \in \mathsf{R}^{L-1}$ the vector consisting of the first $L - 1$ components of Y.
  (f) We set $v^2 = \pi_1^2 + \ldots + \pi_r^2$, where π_i is the last component of the vector P_i $(i = 1, \ldots, r)$. As v^2 is the squared cosine of the angle between the vector $\mathbf{e}_L$ and the linear space $\mathcal{L}_r$, it is called the *verticality coefficient* of $\mathcal{L}_r$. Since $\mathbf{e}_L \notin \mathcal{L}_r$, $v^2 < 1$.

The following statement is fundamental.

Proposition 3.1 *In the notation above, the last component y_L of any vector $Y = (y_1, \ldots, y_L)^\mathsf{T} \in \mathcal{L}_r$ is a linear combination of the first components $y_1, \ldots, y_{L-1}$:*

$$y_L = a_1 y_{L-1} + a_2 y_{L-2} + \ldots + a_{L-1} y_1,$$

where the vector $R = (a_{L-1}, \ldots, a_1)^\mathsf{T}$ can be expressed as

$$R = \frac{1}{1 - v^2} \sum_{i=1}^{r} \pi_i \underline{P_i} \tag{3.1}$$

and does not depend on the choice of the basis $P_1, \ldots, P_r$ in the linear space $\mathcal{L}_r$.

Proof follows from the fact that the formula (3.1) is a particular case of (3.10) below with $n = L, m = r$ and $\mathfrak{Q} = \{L\}$. Another proof of Proposition 3.1 is contained in the proof of [21, Theorem 5.2].

3.1.2 Formal Description of the Algorithms

Below we write down the algorithms of SSA forecasting; see [24, Sect. 3.2.2] and [25] containing the R codes for the calls of the corresponding functions of the RSSA package.

3.1.2.1 Recurrent Forecasting

In the above notation, the *recurrent forecasting algorithm* (briefly, *R-forecasting*) can be formulated as follows.

Algorithm of R-forecasting.

1. The time series $\mathbb{Y}_{N+M} = (y_1, \ldots, y_{N+M})$ is defined by

$$
y_i = \begin{cases} \widetilde{x}_i & \text{for } i = 1, \ldots, N, \\ \sum_{j=1}^{L-1} a_j y_{i-j} & \text{for } i = N+1, \ldots, N+M. \end{cases} \tag{3.2}
$$

2. The numbers $y_{N+1}, \ldots, y_{N+M}$ form the M terms of the recurrent forecast.

This yields that R-forecasting is performed by the direct use of the LRR with coefficients $\{a_j, j = 1, \ldots, L-1\}$ derived in Proposition 3.1.

Remark 3.1 Let us define the linear operator $\mathcal{P}_{\text{Rec}} : \mathbf{R}^L \mapsto \mathbf{R}^L$ by the formula

$$
\mathcal{P}_{\text{Rec}} Y = \begin{pmatrix} \overline{Y} \\ R^{\mathsf{T}} \overline{Y} \end{pmatrix}. \tag{3.3}
$$

Set

$$
Z_i = \begin{cases} \widetilde{X}_i & \text{for } i = 1, \ldots, K, \\ \mathcal{P}_{\text{Rec}} Z_{i-1} & \text{for } i = K+1, \ldots, K+M. \end{cases} \tag{3.4}
$$

It is easily seen that the matrix $\mathbf{Z} = [Z_1 : \ldots : Z_{K+M}]$ is the trajectory matrix of the series $\mathbb{Y}_{N+M}$. Therefore, (3.4) can be regarded as the vector form of (3.2).

3.1.2.2   Vector Forecasting

The idea of *vector forecasting* (briefly, *V-forecasting*) is as follows. Let us assume that we can continue the sequence of vectors $\widehat{X}_1, \ldots, \widehat{X}_K$ (which belong to the subspace $\mathcal{L}_r$) for M steps so that:

(a) the continuation vectors Z_m $(K < m \le K + M)$ belong to the same subspace $\mathcal{L}_r$;
(b) the matrix $\mathbf{X}_M = [\widehat{X}_1 : \ldots : \widehat{X}_K : Z_{K+1} : \ldots : Z_{K+M}]$ is approximately Hankel.

Then, after obtaining the matrix $\mathbf{X}_M$, we can obtain the forecasted series $\mathbb{Y}_{N+M}$ by means of the diagonal averaging of this matrix.

In addition to the notation introduced above let us bring in some more notation. Consider the matrix

$$\boldsymbol{\Pi} = \underline{\mathbf{V}}\,\underline{\mathbf{V}}^{\mathrm{T}} + (1 - v^2)RR^{\mathrm{T}}, \qquad (3.5)$$

where $\underline{\mathbf{V}} = [\underline{P_1} : \ldots : \underline{P_r}]$. The matrix $\boldsymbol{\Pi}$ is the matrix of the linear operator that performs the orthogonal projection $\mathsf{R}^{L-1} \mapsto \mathcal{L}_r$, where $\mathcal{L}_r = \mathrm{span}(\underline{P_1}, \ldots, \underline{P_r})$; note that this matrix $\boldsymbol{\Pi}$ is a particular case of the matrix defined in (3.11) with $m = r$, $n = L$ and $\mathcal{Q} = \{L\}$. Finally, we define the linear operator $\mathcal{P}_{\mathrm{Vec}} : \mathsf{R}^L \mapsto \mathcal{L}_r$ by the formula

$$\mathcal{P}_{\mathrm{Vec}} Y = \begin{pmatrix} \boldsymbol{\Pi}\overline{Y} \\ R^{\mathrm{T}}\overline{Y} \end{pmatrix}. \qquad (3.6)$$

Algorithm of V-forecasting.

1. In the notation above, define the vectors Z_i as follows:

$$Z_i = \begin{cases} \widehat{X}_i & \text{for } i = 1, \ldots, K, \\ \mathcal{P}_{\mathrm{Vec}} Z_{i-1} & \text{for } i = K+1, \ldots, K+M+L-1. \end{cases} \qquad (3.7)$$

2. By constructing the matrix $\mathbf{Z} = [Z_1 : \ldots : Z_{K+M+L-1}]$ and making its diagonal averaging we obtain the series $y_1, \ldots, y_{N+M+L-1}$.
3. The numbers $y_{N+1}, \ldots, y_{N+M}$ form the M terms of the vector forecast.

Remark 3.2   Note that in order to get M forecast terms, the vector forecasting procedure performs $M+L-1$ steps. The aim is the permanence of the forecast under variations in M: the M-step forecast ought to coincide with the first M values of the forecast for $M+1$ or more steps. In view of the definition of the diagonal averaging, to achieve this we have to make $L-1$ extra steps.

3.1.2.3   Simultaneous Forecasting

R-forecasting is based on the fact that the last coordinate of any vector in the subspace $\mathcal{L}_r$ is determined by its first $L - 1$ coordinates. The idea of the *simultaneous forecast-*

ing algorithm is based on the following relation: under some additional conditions, the last M coordinates of any vector in $\mathcal{L}_r$ can be expressed through its first $L - M$ coordinates. Certainly, $L - M$ should be larger than r and therefore $M < L - r$.

Let span$(\mathbf{e}_i, i = L - M + 1, \ldots, L) \cap \mathcal{L}_r = \{\mathbf{0}\}$. For a vector $Y \in \mathcal{L}_r$, denote $Y_1 = (y_1, \ldots, y_{L-M})^{\mathrm{T}}$ and $Y_2 = (y_{L-M+1}, \ldots, y_L)^{\mathrm{T}}$. Then $Y_2 = \mathbf{R}Y_1$, where the matrix $\mathbf{R}$ is defined by (3.10) below with $n = L$, $m = r$ and $\mathfrak{Q} = \{L - M + 1, \ldots, L\}$.

Algorithm of simultaneous forecasting.

1. In the notation above, define the time series $\mathbb{Y}_{N+M} = (y_1, \ldots, y_{N+M})$ by

$$
\begin{aligned}
y_i &= \widetilde{x}_i \quad \text{for} \quad i = 1, \ldots, N, \\
(y_{N+1}, \ldots, y_{N+M})^{\mathrm{T}} &= \mathbf{R}(y_{N-(L-M)+1}, \ldots, y_N)^{\mathrm{T}}.
\end{aligned}
\tag{3.8}
$$

2. The numbers $y_{N+1}, \ldots, y_{N+M}$ form the M terms of the simultaneous forecast.

Remark 3.3 The algorithm formulated above is an analogue of the algorithm of R-forecasting, since $\mathbf{R}$ in (3.8) is applied to the reconstructed series. An analogue of V-forecasting can also be considered.

3.1.3 SSA Forecasting Algorithms: Similarities and Dissimilarities

If $\mathcal{L}_r$ is spanned by certain eigenvectors obtained from the SVD of the trajectory matrix of the series $\mathbb{X}_N$, then the corresponding forecasting algorithm will be called *Basic SSA forecasting algorithm*.

Let us return to Basic SSA and assume that our aim is to extract an additive component $\mathbb{X}_N^{(1)}$ from a series $\mathbb{X}_N$. For an appropriate window length L, we obtain the SVD of the trajectory matrix of the series $\mathbb{X}_N$ and select the eigentriples $(\sqrt{\lambda_i}, U_i, V_i)$, $i \in I$, corresponding to $\mathbb{X}_N^{(1)}$. Then we obtain the resultant matrix

$$
\mathbf{X}_I = \sum_{i \in I} \sqrt{\lambda_i} U_i V_i^{\mathrm{T}}
$$

and, after the diagonal averaging, we obtain the reconstructed series $\widetilde{\mathbb{X}}_N^{(1)}$ that estimates $\mathbb{X}_N^{(1)}$.

The columns $\widehat{X}_1, \ldots, \widehat{X}_K$ of the resultant matrix $\mathbf{X}_I$ belong to the linear space $\mathcal{L}_r = \mathrm{span}(U_i, i \in I)$. If $\mathbb{X}_N^{(1)}$ is strongly separable from $\mathbb{X}_N^{(2)} \stackrel{\mathrm{def}}{=} \mathbb{X}_N - \mathbb{X}_N^{(1)}$, then $\mathcal{L}_r$ coincides with $\mathcal{X}^{(L,1)}$ (the trajectory space of the series $\mathbb{X}_N^{(1)}$) and $\mathbf{X}_I$ is a Hankel matrix (in this case, $\mathbf{X}_I$ is the trajectory matrix of the series $\mathbb{X}_N^{(1)}$). Then the recurrent, vector and simultaneous forecasts coincide and the resulting procedure could be called the *exact continuation* of $\mathbb{X}_N^{(1)}$. More precisely, in this situation the matrix Π

is the identity matrix, and (3.6) coincides with (3.3). Furthermore, the matrix **Z** has Hankel structure and the diagonal averaging does not change the matrix elements.

If $\mathbb{X}_N^{(1)}$ and $\mathbb{X}_N^{(2)}$ are approximately strongly separable, then $\mathcal{L}_r$ is close to $\mathfrak{X}^{(L,1)}$ and $\mathbf{X}_I$ is approximately a Hankel matrix.

If there is no exact separability, then different forecasting algorithms usually give different results. Let us describe the difference between them. Since the recurrent and vector forecasting algorithms are more conventional and have less limitations, we shall concentrate on the recurrent and vector forecasting algorithms only.

- In a typical situation, there is no time series such that the linear space $\mathcal{L}_r$ (for $r < L-1$) is its trajectory space [21, Proposition 5.6], and therefore this space cannot be the trajectory space of the series to be forecasted. The R-forecasting method uses $\mathcal{L}_r$ to obtain the LRR of the forecasting series. The V-forecasting procedure tries to perform the L-continuation of the series in $\mathcal{L}_r$: any vector $Z_{i+1} = \mathcal{P}_{\text{Vec}} Z_i$ belongs to $\mathcal{L}_r$, and Z_{i+1} is as close as possible to $\overline{Z_i}$. The last component of Z_{i+1} is obtained from $\overline{Z_{i+1}}$ by the same LRR as used in R-forecasting.
- Both forecasting methods have two general stages: the diagonal averaging and continuation. For R-forecasting, the diagonal averaging is used to obtain the reconstructed series, and continuation is performed by applying the LRR. In V-forecasting, these two stages are used in the reverse order; first, vector continuation in $\mathcal{L}_r$ is performed and then the diagonal averaging gives the forecast.
- If there is no exact separability it is hard to compare the recurrent and vector forecasting methods theoretically. Closeness of two forecasts obtained by two different algorithms can be used as an argument in favour of forecasting stability.
- R-forecasting is simpler to interpret in view of the link between LRRs and their characteristic polynomials, see Sect. 3.2. On the other hand, numerical study demonstrates that V-forecasting is typically more 'conservative' (or less 'radical') when R-forecasting may exhibit either rapid increase or decrease.
- Only very few years ago, V-forecasting was considered as having a larger computational cost than R-forecasting; however, due to the current efficient implementation in Rssa [23], V-forecasting has become even slightly faster than R-forecasting.

Remark 3.4 Forecasting algorithms described in Sect. 3.1 are based on the estimation of the trajectory subspace of the forecasted component. In addition to Basic SSA, there are other methods of estimation of the trajectory subspace. For example, if the subspace is estimated by Toeplitz SSA, we obtain Toeplitz SSA forecasting algorithms. We may wish to use SSA with centering for estimating the subspace; in this case, we arrive at corresponding modifications of SSA forecasting with centering, see [21, Sect. 2.3.3].

3.1.4 Appendix: Vectors in a Subspace

In this section, we formulate two technical results that provide the theoretical ground for both forecasting and filling in methods. For proofs and details, we refer to [16].

Consider the Euclidean space R^n. Define $J_n = \{1, \ldots, n\}$ and denote by $\Omega = \{i_1, \ldots, i_s\} \subset J_n$ an ordered set, $|\Omega| = s$. Let $\mathbf{I}_s$ denote the unit $s \times s$ matrix. We define a *restriction of a vector* $X = (x_1, \ldots, x_n)^{\mathsf{T}} \in \mathsf{R}^n$ onto a set of indices $\Omega = \{i_1, \ldots, i_s\}$ as the vector $X\big|_\Omega = (x_{i_1}, \ldots, x_{i_s})^{\mathsf{T}} \in \mathsf{R}^s$. *The restriction of a matrix* onto a set of indices is the matrix consisting of restrictions of its column vectors onto this set.

The restriction of a q-dimensional subspace $\mathcal{L}_q$ onto a set of indices Ω is the space spanned by restrictions of all vectors of $\mathcal{L}_q$ onto this set; the restricted space will be denoted by $\mathcal{L}_q\big|_\Omega$. It is easy to prove that for any basis $\{H_i\}_{i=1}^q$ of the subspace $\mathcal{L}_q$, the equality $\mathcal{L}_q\big|_\Omega = \mathrm{span}\big(H_1\big|_\Omega, \ldots, H_q\big|_\Omega\big)$ holds.

3.1.4.1 Filling in Vector Coordinates in the Subspace

Consider an m-dimensional subspace $\mathcal{L}_m \subset \mathsf{R}^n$ with $m < n$. Denote by $\{P_k\}_{k=1}^m$ an orthonormal basis in $\mathcal{L}_m$ and define the matrix $\mathbf{P} = [P_1 : \ldots : P_m]$. Fix an ordered set of indices $\Omega = \{i_1, \ldots, i_s\}$ with $s = |\Omega| \le n - m$.

First, note that the following conditions are equivalent (it follows from [16, Lemma 2.1]): (1) for any $Y \in \mathcal{L}_m\big|_{J_n\setminus\Omega}$ there exists an unique vector $X \in \mathcal{L}_m$ such that $X\big|_{J_n\setminus\Omega} = Y$, (2) the matrix $\mathbf{I}_s - \mathbf{P}\big|_\Omega\big(\mathbf{P}\big|_\Omega\big)^{\mathsf{T}}$ is non-singular, and (3) $\mathrm{span}(\mathbf{e}_i, i \in \Omega) \cap \mathcal{L}_m = \{\mathbf{0}_n\}$. Either of these conditions can be considered as a condition of unique filling in of the missing vector components with indices from Ω.

Proposition 3.2 *Let the matrix* $\mathbf{I}_s - \mathbf{P}\big|_\Omega\big(\mathbf{P}\big|_\Omega\big)^{\mathsf{T}}$ *be non-singular. Then for any vector* $X \in \mathcal{L}_m$ *we have*

$$X\big|_\Omega = \mathbf{R}\, X\big|_{J_n\setminus\Omega}, \tag{3.9}$$

where

$$\mathbf{R} = \Big(\mathbf{I}_s - \mathbf{P}\big|_\Omega\big(\mathbf{P}\big|_\Omega\big)^{\mathsf{T}}\Big)^{-1} \mathbf{P}\big|_\Omega\big(\mathbf{P}\big|_{J_n\setminus\Omega}\big)^{\mathsf{T}}. \tag{3.10}$$

3.1.4.2 Projection Operator

Let $Y \in \mathsf{R}^n$ and $Z = Y\big|_{J_n\setminus\Omega} \in \mathsf{R}^{n-s}$. Generally, $Z \notin \mathcal{L}_m\big|_{J_n\setminus\Omega}$. However, for applying formula (3.9) to obtain the vector from $\mathcal{L}_m$, it is necessary that $Z \in \mathcal{L}_m\big|_{J_n\setminus\Omega}$. The orthogonal projector $\mathsf{R}^{n-s} \to \mathcal{L}_m\big|_{J_n\setminus\Omega}$ transfers Z to $\mathcal{L}_m\big|_{J_n\setminus\Omega}$.

Set $\mathbf{V} = \mathbf{P}\big|_{J_n\setminus\Omega}$ and $\mathbf{W} = \mathbf{P}\big|_\Omega$ for the convenience of notation. The matrix of the projection operator $\mathbf{\Pi}_{J_n\setminus\Omega}$ can be derived as follows.

Proposition 3.3 *Assume that the matrix* $\mathbf{I}_s - \mathbf{W}\mathbf{W}^{\mathsf{T}}$ *is nonsingular. Then the matrix of the orthogonal projection operator* $\mathbf{\Pi}_{J_n\setminus\Omega}$ *has the form*

$$\Pi_{J_n \backslash Q} = \mathbf{V}\mathbf{V}^{\mathrm{T}} + \mathbf{V}\mathbf{W}^{\mathrm{T}}(\mathbf{I}_s - \mathbf{W}\mathbf{W}^{\mathrm{T}})^{-1}\mathbf{W}\mathbf{V}^{\mathrm{T}}. \tag{3.11}$$

3.2 LRR and Associated Characteristic Polynomials

The theory of the linear recurrence relations and associated characteristic polynomials is well known (for example, see [11, Chap. V, Sect. 4]). Here we provide a short survey of the results which are most essential for understanding SSA forecasting.

Definition 3.1 A time series $\mathbb{S}_N = \{s_i\}_{i=1}^N$ is *governed by an LRR*, if there exist $a_1, \ldots, a_t$ such that

$$s_{i+t} = \sum_{k=1}^{t} a_k s_{i+t-k}, \ 1 \le i \le N - t, \ a_t \ne 0, \ t < N. \tag{3.12}$$

The number t is called the order of the LRR, $a_1, \ldots, a_t$ are the coefficients of the LRR. If $t = r$ is the minimal order of an LRR that governs the time series $\mathbb{S}_N$, then the corresponding LRR is called *minimal* and we say that the time series $\mathbb{S}_N$ has *finite-difference dimension* r.

Note that if the minimal LRR governing the signal $\mathbb{S}_N$ has order r with $r < N/2$, then $\mathbb{S}_N$ has rank r (see Sect. 2.3.1.2 for the definition of the series of finite rank).

Definition 3.2 A polynomial $P_t(\mu) = \mu^t - \sum_{k=1}^{t} a_k \mu^{t-k}$ is called a *characteristic polynomial* of the LRR (3.12).

Let the time series $\mathbb{S}_\infty = (s_1, \ldots, s_n, \ldots)$ satisfy the LRR (3.12) with $a_t \ne 0$ and $i \ge 1$. Consider the characteristic polynomial of the LRR (3.12) and denote its different (complex) roots by $\mu_1, \ldots, \mu_p$ with $1 \le p \le t$. All these roots are non-zero as $a_t \ne 0$. Let the multiplicity of the root μ_m be k_m, where $1 \le m \le p$ and $k_1 + \ldots + k_p = t$. The following well-known result (see e.g. [21, Theorem 5.3] or [27]) provides an explicit form for the series which satisfies the LRR.

Theorem 3.1 *The time series* $\mathbb{S}_\infty = (s_1, \ldots, s_n, \ldots)$ *satisfies the LRR* (3.12) *for all* $i \ge 0$ *if and only if*

$$s_n = \sum_{m=1}^{p} \left(\sum_{j=0}^{k_m-1} c_{mj} n^j \right) \mu_m^n, \tag{3.13}$$

where the complex coefficients c_{mj} depend on the first t points $s_1, \ldots, s_t$.

For the real-valued time series, Theorem 3.1 implies that the class of time series governed by the LRRs consists of sums of products of polynomials, exponentials and sinusoids.

3.2.1 Roots of the Characteristic Polynomials

Let the series $\mathbb{S}_N = (s_1, \ldots, s_N)$ be governed by an LRR (3.12) of order t. Let $\mu_1, \ldots, \mu_p$ be pair-wise different (complex) roots of the characteristic polynomial $P_t(\mu)$. As $a_t \neq 0$, all these roots are not equal to zero. We also have $k_1 + \ldots + k_p = t$, where k_m are the multiplicities of the roots μ_m ($m = 1, \ldots, p$).

Denote $s_n(m, j) = n^j \mu_m^n$ for $1 \leq m \leq p$ and $0 \leq j \leq k_m - 1$. Theorem 3.1 tells us that the general solution of the equation (3.12) is

$$s_n = \sum_{m=1}^{p} \sum_{j=0}^{k_m-1} c_{mj} s_n(m, j) \tag{3.14}$$

with certain complex c_{mj}. The coefficients c_{mj} are defined by $s_1, \ldots, s_t$, the first t elements of the series $\mathbb{S}_N$.

Thus, each root μ_m produces a component

$$s_n^{(m)} = \sum_{j=0}^{k_m-1} c_{mj} s_n(m, j) \tag{3.15}$$

of the series $\mathbb{S}_N$. Let us fix m and consider the m-th component in the case $k_m = 1$, which is the main case in practice. Set $\mu_m = \rho e^{i2\pi\omega}$, $\omega \in (-1/2, 1/2]$, where $\rho > 0$ is the modulus (absolute value) of the root and $2\pi\omega$ is its polar angle.

If ω is either 0 or $1/2$, then μ_m is a real root of the polynomial $P_t(\mu)$ and the series component $s_n^{(m)}$ is real and is equal to $c_{m0}\mu_m^n$. This means that $s_n^{(m)} = A\rho^n$ for positive μ_m and $s_n^{(m)} = A(-1)^n \rho^n = A\rho^n \cos(\pi n)$ for negative μ_m. This last case corresponds to the exponentially modulated saw-tooth sequence.

All other values of ω lead to complex μ_m. In this case, P_t has a complex conjugate root $\mu_l = \rho e^{-i2\pi\omega}$ of the same multiplicity $k_l = 1$. We thus can assume $0 < \omega < 1/2$ and describe a pair of conjugate roots by the pair of real numbers (ρ, ω) with $\rho > 0$ and $\omega \in (0, 1/2)$.

By adding up the components $s_n^{(m)}$ and $s_n^{(l)}$ corresponding to these conjugate roots we obtain the real series $A\rho^n \cos(2\pi\omega n + \varphi)$, with A and φ expressed in terms of c_{m0} and c_{l0}. The frequency ω can be expessed in the form of the period $T = 1/\omega$ and vice versa.

The asymptotic behaviour of $s_n^{(m)}$ mainly depends on $\rho = |\mu_m|$. Let us consider the simplest case $k_m = 1$ as above. If $\rho < 1$, then $s_n^{(m)}$ rapidly tends to zero and asymptotically has no influence on the whole series (3.14). Alternatively, the root with $\rho > 1$ and $|c_{m0}| \neq 0$ leads to a rapid increase of $|s_n|$ (at least, of a certain subsequence of $\{|s_n|\}$).

Let r be the finite-difference dimension of a series $\mathbb{S}_N$. Then the characteristic polynomial of the minimal LRR of $\mathbb{S}_N$ has order r and it has r roots. The same series satisfies many other LRRs of dimensions $t > r$. Consider any such LRR (3.12) with $t > r$. The characteristic polynomial $P_t(\mu)$ of the LRR (3.12) has t roots with r

roots (we call them the *main roots*) coinciding with the roots of the minimal LRR. The other $t - r$ roots are *extraneous*: in view of the uniqueness of the representation (3.15), the coefficients c_{mj} corresponding to these roots are equal to zero. However, the LRR (3.12) governs a wider class of series than the minimal LRR.

Since the roots of the characteristic polynomial specify its coefficients uniquely, they also determine the corresponding LRR. Consequently, by removing the extraneous roots of the characteristic polynomial $P_t(\mu)$, corresponding to the LRR (3.12), we can obtain the polynomial describing the minimal LRR of the series.

Example 3.1 (*Annual periodicity*) Assume that the series $\mathbb{S}_N$ has zero mean and period 12. Then it can be expressed as a sum of six harmonics:

$$
s_n = \sum_{k=1}^{5} c_k \cos(2\pi nk/12 + \varphi_k) + c_6 \cos(\pi n). \tag{3.16}
$$

Under the condition $c_k \neq 0$ for $k = 1, \ldots, 6$, the series has finite-difference dimension 11. In other words, the characteristic polynomial of the minimal LRR governing the series (3.16) has 11 roots. All these roots have modulus 1. The real root -1 corresponds to the last term in (3.16). The harmonic term with frequency $\omega_k = k/12$ $(k = 1, \ldots, 5)$ generates two complex conjugate roots $\exp(\pm i 2\pi k/12)$, which have polar angles $\pm 2\pi k/12$. $\qquad \Box$

3.2.2   Min-Norm LRR

Consider a time series $\mathbb{S}_N$ of rank r governed by an LRR. Let L be the window length ($r < \min(L, K)$, $K = N - L + 1$), $\mathbf{S}$ be the trajectory matrix of $\mathbb{S}_N$, $\mathcal{S}$ be its trajectory space, $P_1, \ldots, P_r$ form an orthonormal basis of $\mathcal{S}$ and $\mathcal{S}^\perp$ be the orthogonal complement to $\mathcal{S}$. Denote $A = (a_{L-1}, \ldots, a_1, -1)^{\mathrm{T}} \in \mathcal{S}^\perp$, $a_{L-1} \neq 0$. Then the time series $\mathcal{S}$ satisfies the LRR

$$
s_{i+(L-1)} = \sum_{k=1}^{L-1} a_k s_{i+(L-1)-k}, \ 1 \le i \le K. \tag{3.17}
$$

Conversely, if a time series is governed by an LRR (3.17), then the LRR coefficients $B = (a_{L-1}, \ldots, a_1)^{\mathrm{T}}$ complemented with -1 yield the vector $\begin{pmatrix} B \\ -1 \end{pmatrix} \in \mathcal{S}^\perp$. Note that any LRR that governs a time series can be treated as a forward linear prediction. In addition, if we consider a vector in $\mathcal{S}^\perp$ with -1 as the first coordinate, then we obtain the so-called backward linear prediction [46].

For any matrix $\mathbf{A}$, we denote by $\underline{\mathbf{A}}$ the matrix $\mathbf{A}$ with the last row removed and by $\overline{\mathbf{A}}$ the matrix $\mathbf{A}$ without the first row.

From the viewpoint of prediction, the LRR governing a time series of rank r has coefficients derived from the condition $\underline{S}^T B = (s_L, \ldots, s_N)^T$. This system of linear equations may have several solutions, since the vector $(s_L, \ldots, s_N)^T$ belongs to the column space of the matrix $\underline{S}^T$. It is well-known that the least-squares solution expressed by the Moore–Penrose pseudo-inverse to $\underline{S}^T$ yields the vector B with minimum norm (the solution for the method of total least squares coincides with it). It can be shown that this minimum-norm solution B_{LS} can be expressed as

$$B_{LS} = (a_{L-1}, \ldots, a_1)^T = \frac{1}{1 - v^2} \sum_{i=1}^{r} \pi_i \, \underline{P_i}, \tag{3.18}$$

where π_i are the last coordinates of P_i and $v^2 = \sum_{i=1}^{r} \pi_i^2$.

Thus, one of the vectors from $S^\perp$, which equals $A_{LS} = \begin{pmatrix} B_{LS} \\ -1 \end{pmatrix}$, has a special significance and the corresponding LRR is called the *min-norm LRR*; it provides the min-norm (forward) prediction. Similarly, we can derive a relation for the min-norm backward prediction.

It is shown in [21, Proposition 5.5] and [31] that the forward min-norm prediction vector A_{LS} is the normalized (so that its last coordinate is equal to -1) projection of the L-th coordinate vector $\mathbf{e}_L$ on $S^\perp$, the orthogonal complement to the signal subspace. Therefore, the min-norm prediction vector depends on the signal subspace only.

The following property demonstrates the importance of the minimum norm of the LRR coefficients for noise reduction.

Proposition 3.4 *Let $\mathbb{X}_N = \mathbb{S}_N + \mathbb{P}_N$, where $\mathbb{P}_N$ is a stationary white noise with zero mean and variance σ^2, X and S be L-lagged vectors of $\mathbb{X}_N$ and $\mathbb{S}_N$ correspondingly and $C \in \mathsf{R}^{L-1}$. Then for $x = C^T S$ and $\tilde{x} = C^T X$, we have $\mathsf{E}\tilde{x} = x$ and $\mathsf{D}\tilde{x} = \|C\|^2 \sigma^2$.*

The proof directly follows from the equality $\mathsf{D} \sum_{i=1}^{L-1} c_i (y_i + \varepsilon_i) = \mathsf{D} \sum_{i=1}^{L-1} c_i \varepsilon_i = \|C\|^2 \sigma^2$, where $C = (c_1, \ldots, c_{L-1})^T$ and $\varepsilon_i, i = 1, \ldots, L-1$ are i.i.d. random variables with zero mean and variance σ^2.

If $X = X_K$ is the last lagged vector of $\mathbb{S}_N$, then $\tilde{x} = C^T \overline{X}_K$ can be considered as a forecasting formula applied to a noisy signal and $\|C\|^2$ regulates the variance of this forecast.

The following property of the min-norm LRR, which was derived in [32], is extremely important for forecasting: all extraneous roots of the min-norm LRR lie inside the unit circle of the complex plane. Example 3.2, where the min-norm LRR is used, illustrates this property giving us hope that in the case of real-life series (when both the min-norm LRR and the related initial data are perturbed) the terms related to the extraneous roots in (3.13) only slightly influence the forecast. Moreover, bearing in mind the results concerning the distribution of the extraneous roots (see [38, 47]), we can expect that the extraneous summands cancel each other out.

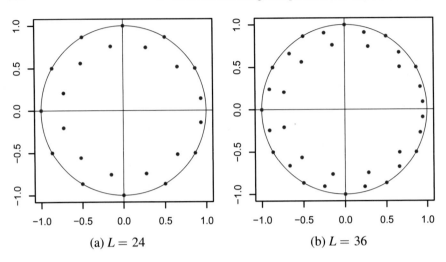

Fig. 3.1 Annual periodicity: main and extraneous roots

Example 3.2 (*Annual periodicity and extraneous roots*) Let us consider the series
(3.16) from Example 3.1 and the min-norm LRR, which is not minimal. Let N be
large enough. If we select certain $L \geq 12$ and take $r = 11$ and $\mathcal{L}_r = \mathcal{S}(\mathbb{S}_N)$, then the
vector $R = (a_{L-1}, \ldots, a_1)^{\mathrm{T}}$ defined in (3.18) produces the LRR (3.17), which is not
minimal but governs the series (3.16).

Let us take $c_i = i - 1$, $\varphi_1 = \ldots = \varphi_5 = 0$ and $L = 24$, 36. The roots of the
characteristic polynomials of the LRR (3.17) are depicted in Fig. 3.1. We can see
that the main 11 roots of the polynomial compose 11 of 12 vertices of a regular
dodecagon and lie on the unit circle in the complex plane. Twelve ($L = 24$) and
twenty four ($L = 36$) extraneous roots have smaller moduli.   □

Remark 3.5 Note that the min-norm LRR forms the basis for the SSA forecasting
methods introduced in Sect. 3.1 (see [21, Sect. 2.1]). In particular, R-forecasting uses
the estimated min-norm LRR for forecasting: compare the formulas (3.1) and (3.18)
for the coefficients of the LRRs.

3.3 Recurrent Forecasting as Approximate Continuation

Exact continuation is hardly practically significant. Indeed, it seems unwise to assume
that a real-life time series is governed by some LRR of relatively small dimension.
Therefore, we need to consider approximate continuation; it is of much greater importance in practice than exact continuation. In this section we consider approximate
continuation with the help of recurrent forecasting. Most discussions are also relevant
for other SSA forecasting algorithms.

3.3.1 Approximate Separability and Forecasting Errors

Let $\mathbb{X}_N = \mathbb{X}_N^{(1)} + \mathbb{X}_N^{(2)}$ and suppose that the time series $\mathbb{X}_N^{(1)}$ admits a recurrent continuation. Denote by d the dimension of the minimal recurrence relation governing $\mathbb{X}_N^{(1)}$. If $d < \min(L, N - L + 1)$, then $d = \operatorname{rank}_L(\mathbb{X}_N^{(1)})$.

If $\mathbb{X}_N^{(1)}$ and $\mathbb{X}_N^{(2)}$ are strongly separable for some window length L, then the trajectory space of $\mathbb{X}_N^{(1)}$ can be found and we can perform recurrent continuation of the series $\mathbb{X}_N^{(1)}$ by the method described in Sect. 3.1.2.1. We now assume that $\mathbb{X}_N^{(1)}$ and $\mathbb{X}_N^{(2)}$ are approximately strongly separable and discuss the problem of approximate continuation (forecasting) of the series $\mathbb{X}_N^{(1)}$ in the subspace $\mathcal{L}_r$. The choice of $\mathcal{L}_r$ is described in Sect. 3.1.3. If the choice is proper, $r = d$.

The series of forecasts y_n $(n > N)$ defined by (3.2) generally does not coincide with the recurrent continuation of the series $\mathbb{X}_N^{(1)}$. The deviation between these two series makes the forecasting error. This error has two origins. The main origin is the difference between the linear space $\mathcal{L}_r$ and $\mathcal{X}^{(L,1)}$, the trajectory space of the series $\mathbb{X}_N^{(1)}$ (some inequalities connecting the perturbation of the LRR (3.2) with that of $\mathcal{X}^{(L,1)}$ are derived in [36], see also the end of Sect. 2.3.3 for a related discussion). Since the LRR (3.2) is produced by the vector R and the latter is strongly related to the space $\mathcal{L}_r$, the discrepancy between $\mathcal{L}_r$ and $\mathcal{X}^{(L,1)}$ produces an error in the LRR governing the series of forecasts. In particular, the finite-difference dimension of the series of forecasts y_n $(n > N)$ is generally larger than r.

The other origin of the forecasting error lies in the initial data used to build the forecast. In the case of recurrent continuation, the initial data is $x_{N-L+2}^{(1)}, \ldots, x_N^{(1)}$, where $x_n^{(1)}$ is the n-th term of the series $\mathbb{X}_N^{(1)}$. In the Basic SSA R-forecasting algorithm, the initial data consists of the last $L-1$ terms $y_{N-L+2}, \ldots, y_N$ of the reconstructed series. Since generally $x_n^{(1)} \neq y_n$, the initial data used in LRR is a source of forecasting errors. The splitting of the whole error into two parts is investigated in [14] by simulations. For L close to $N/2$, these parts are comparable while for small L the contribution of the error caused by the wrong reconstruction is larger.

On the other hand, if the quality of approximate separability of $\mathbb{X}_N^{(1)}$ and $\mathbb{X}_N^{(2)}$ is rather good and we select the proper eigentriples associated with $\mathbb{X}^{(1)}$, then we can expect that the linear spaces $\mathcal{L}_r$ and $\mathcal{X}^{(L,1)}$ are close. Therefore, the coefficients in the LRR (3.2) are expected to be close to those of the LRR governing the recurrent continuation of the series $\mathbb{X}_N^{(1)}$. Similarly, approximate separability implies that the reconstructed series y_n is close to $x_n^{(1)}$ and therefore the error due to the imprecision of the initial data used for forecasting is also small. As a result, in this case we can expect that the Basic SSA R-forecasting procedure provides a reasonably accurate approximation to the recurrent continuation of $\mathbb{X}_N^{(1)}$, at least in the first few steps.

Remark 3.6 Since the forecasting procedure contains two generally unrelated parts, namely, estimation of the LRR and estimation of the reconstruction, we can modify these two parts of the algorithm separately. For example, for forecasting a signal, the LRR can be applied to the initial time series if the last points of the reconstruction are expected to be biased. Another modification of the forecasting procedure is considered in [14]; it is based on the use of different window lengths for estimation of the LRR and for reconstruction of the time series.

3.3.2   Approximate Continuation and Characteristic Polynomials

In this section, we continue discussing the errors of separability and forecasting. The discrepancy between $\mathcal{L}_r$ and $\mathfrak{X}^{(L,1)}$ will be assessed in terms of the characteristic polynomials.

We have three LRRs: (i) the minimal LRR of order r governing the time series $\mathbb{X}_N^{(1)}$, (ii) the continuation LRR of order $L-1$, which governs $\mathbb{X}_N^{(1)}$ but also produces $L-r-1$ extraneous roots in its characteristic polynomial P_{L-1}, and (iii) the forecasting min-norm LRR governing the series of forecasts y_n $(n > N)$.

The characteristic polynomial $P_{L-1}^{(x)}$ of the forecasting LRR and continuation polynomial P_{L-1} have $L-1$ roots. If $\mathcal{L}_r$ and $\mathfrak{X}^{(L,1)}$ are close, then the coefficients of the continuation and forecasting recurrence relations must be close too. Therefore, all simple roots of the forecasting characteristic polynomial $P_{L-1}^{(x)}$ must be close to the roots of the continuation polynomial P_{L-1}. The roots μ_m with multiplicities $k_m > 1$ could be perturbed in a more complex manner.

Example 3.3 (*Perturbation of the multiple roots*) Let us consider the series $\mathbb{X}_N$ with

$$x_n = (A + 0.1\,n) + \sin(2\pi n/10), \quad n = 0, \ldots, 199.$$

Evidently, $\mathbb{X}_N = \mathbb{X}_N^{(1)} + \mathbb{X}_N^{(2)}$ with the linear series $\mathbb{X}_N^{(1)}$ defined by $x_{n+1}^{(1)} = A + 0.1\,n$ and the harmonic series $\mathbb{X}_N^{(2)}$ corresponding to $x_{n+1}^{(2)} = \sin(2\pi n/10)$.

The series $\mathbb{X}_N$ has rank 4 and is governed by the minimal LRR of order 4. Therefore, any LRR governing $\mathbb{X}_N$ produces a characteristic polynomial with four main roots. These main roots do not depend on A; the linear part of the series generates one real root $\mu = 1$ of multiplicity 2, while the harmonic series corresponds to two complex conjugate roots $\rho e^{\pm i2\pi\omega}$ with modulus $\rho = 1$ and frequency $\omega = 0.1$.

Our aim is to forecast the series $\mathbb{X}_N^{(1)}$ for $A = 0$ and $A = 50$ with the help of the Basic SSA R-forecasting algorithm. In both cases, we take the window length $L = 100$ and choose the eigentriples that correspond to the linear part of the initial time series $\mathbb{X}_N$. (For $A = 0$ we take the two leading eigentriples, while for $A = 50$ the appropriate eigentriples have the ordinal numbers 1 and 4.) Since the series $\mathbb{X}_N^{(1)}$ and $\mathbb{X}_N^{(2)}$ are not exactly separable for any A and any choice of L, we deal with approximate separability. The forecasting polynomials $P_{L-1}^{(x)}$ with $A = 0$ and $A = 50$ demonstrate different splitting of the double root $\mu = 1$ into two simple ones. For $A = 0$ there appear two complex conjugate roots with $\rho = 1.002$ and $\omega = 0.0008$, while in the case $A = 50$ we obtain two real roots equal to 1.001 and 0.997. All extraneous roots are smaller than 0.986. This means that for $A = 0$ the linear series $\mathbb{X}_N^{(1)}$ is approximated by a low-frequency harmonic with slightly increasing exponential amplitude. In the case $A = 50$, the approximating series is the sum of two exponentials, one of them is slightly increasing and another one is slightly decreasing. Therefore, we have different long-term forecasting formulas: oscillating for $A = 0$ and exponentially increasing for $A = 50$. For short-term forecasting, this difference is not important. $\qquad\qquad\qquad\qquad\qquad\qquad\qquad\qquad\qquad\qquad\qquad\qquad\quad$ □

Let us consider the part of the forecasting error caused by errors in the initial data, that is, in the reconstruction of the forecasted series component. If the LRR is not minimal ($L > r + 1$), then the corresponding characteristic polynomial P_{L-1} has $L - 1 - r$ extraneous roots. If there is no reconstruction error, then the extraneous roots do not affect the forecast since the coefficients c_{mj} in (3.13) for the corresponding summands are equal to zero. However, if one applies the LRR to the perturbed initial terms, then the extraneous roots start to affect the forecasting results. The extraneous roots of the min-norm LRR lie within the unit circle and their effect on the forecasting decreases for long-term forecasting. Unfortunately, the minimal LRR is not appropriate for forecasting as it is very sensitive to errors in the initial data. Hence the presence of extraneous roots should be taken into account.

In the case of approximate separability, the min-norm LRR is found approximately. As a consequence, the extraneous roots can have absolute values larger than 1. The extraneous roots with moduli greater than 1 are very hazardous, since the extraneous summand μ^n in (3.13), caused by an extraneous root μ with $|\mu| > 1$, grows indefinitely. Therefore, it is important to look at the extraneous roots of the LRR used for forecasting.

If the forecasted series component $\mathbb{X}_N^{(1)}$ is the signal, then the main roots can be called signal roots. Note that the knowledge of extraneous roots should be used both for finding the parametric form (3.13) of the signal (then we should identify the signal roots and remove the extraneous roots) and also for forecasting the signal (then we do not need to know the values of the roots but we would like to have no extraneous roots outside the unit circle).

Since the forecasting LRR is fully determined by the roots of its characteristic polynomial, certain manipulations with the polynomial roots can be performed to modify the R-forecasting procedure.

- Let the main roots of the min-norm LRR of order $L - 1$ be identified or estimated (e.g. by ESPRIT, see Sect. 3.8.2). For example, for a time series with the signal components which are not decreasing, the estimated main roots typically have maximal moduli (since the extraneous roots lie inside the unit circle). Thereby, we obtain the estimated minimal LRR, which is also the min-norm LRR of order r. However, it follows from the definition of the minimum norm that the norm of the coefficients of the minimal LRR is larger than that of the min-norm LRR of order $L - 1$ for $L > r + 1$. Therefore, the forecast by the minimal LRR is more sensitive to errors in the initial data. Simulations demonstrate that in most cases the use of the minimal LRR does not give the most accurate forecast and, moreover, such forecast is often rather unstable.

- A safe way of correcting the LRR is by adjusting the identified main roots when an additional information about the time series is available. For example, if we know that the forecasted oscillations have stationary periodicities with constant amplitudes, then we know that the root moduli are equal to one and therefore the corresponding roots can be substituted with $\mu' = \mu/\|\mu\|$. If there is a periodicity with known period in the time series, then we can correct the arguments of the corresponding roots (for example, to 1/12, 1/6 and so on for a monthly data with seasonality).
- If the main roots have been estimated, then an explicit formula for the time series values in the form (3.13) can be obtained (using estimates of c_{mj} obtained by the least-squares method) and the forecast can be produced by this explicit formula. However, an explicit forecast needs root estimation, whereas R-forecasting does not need root estimation and therefore it is more robust.

3.4   Confidence Bounds for the Forecasts

There are several conventional ways of estimating the accuracy of a forecast. Most of them can be applied for forecasting of the signal in the signal plus noise model.

1. Theoretical confidence intervals can be constructed if the model of time series is known and there are theoretical results about the distribution of the forecast.
2. Bootstrap confidence intervals can be constructed if the model of the signal is estimated in the course of analysis.
3. The accuracy of forecasting can be tested by removal of the last few points and then forecasting their values (so-called *retrospective forecast*). This can be repeated with the cut made at different points.
4. If we are not interested in the retrospective forecast (we really need to forecast the future) and cannot reliably build an SSA model (as well as any other model) then we can use the following approach: we build a large number of SSA forecasts (e.g. using a variety of L and different but reasonable grouping schemes) and compare the forecast values at the horizon we are interested in. If the forecasts are going all over the place then we cannot trust any of them. If however the variability of the constructed forecasts is small then we (at least partly) may trust them; see [39] for details and examples.

If there is a set of possible models, then the model can be chosen by minimizing the forecasting errors. An adjustment taking into account the number of parameters in the models should be made similar to the methods based on characteristics like the Akaike information criterion or by using the degrees-of-freedom adjustments, see discussion in Sect. 2.4.4.3.

There are not enough theoretical results which would help in estimating the accuracy of SSA forecasts theoretically. Below in this section we consider bootstrap confidence intervals in some detail. Since construction of bootstrap confidence intervals is very similar to that of the Monte Carlo confidence intervals, we also consider

Monte Carlo techniques for the investigation of the accuracy of reconstruction and forecasting. Note that by constructing bootstrap confidence intervals for forecasting values we also obtain confidence limits for the reconstructed values.

3.4.1 Monte Carlo and Bootstrap Confidence Intervals

According to the main SSA forecasting assumptions, the component $\mathbb{X}_N^{(1)}$ of the time series $\mathbb{X}_N$ ought to be governed by an LRR of relatively small dimension, and the residual series $\mathbb{X}_N^{(2)} = \mathbb{X}_N - \mathbb{X}_N^{(1)}$ ought to be approximately strongly separable from $\mathbb{X}_N^{(1)}$ for some window length L. In particular, $\mathbb{X}_N^{(1)}$ is assumed to be a finite subseries of an infinite series, which is a recurrent continuation of $\mathbb{X}_N^{(1)}$. These assumptions hold for a wide class of practical series.

To establish confidence bounds for the forecast, we have to apply even stronger assumptions, related not only to $\mathbb{X}_N^{(1)}$, but to $\mathbb{X}_N^{(2)}$ as well. We assume that $\mathbb{X}_N^{(2)}$ is a finite subseries of an infinite random noise series $\mathbb{X}^{(2)}$ that perturbs the signal $\mathbb{X}^{(1)}$.

We only consider Basic SSA R-forecasting method. All other SSA forecasting procedures can be treated analogously.

Let us consider a method of constructing confidence bounds for the signal $\mathbb{X}^{(1)}$ at the moment of time $N + M$. In the unrealistic situation, when we know both the signal $\mathbb{X}^{(1)}$ and the true model of the noise $\mathbb{X}_N^{(2)}$, a direct Monte Carlo simulation can be used to check statistical properties of the forecast value $\widetilde{x}_{N+M}^{(1)}$ relative to the actual value $x_{N+M}^{(1)}$. Indeed, assuming that the rules for the eigentriple selection are fixed, we can simulate Q independent copies $\mathbb{X}_{N,i}^{(2)}$ of the process $\mathbb{X}_N^{(2)}$ and apply the forecasting procedure to Q independent time series $\mathbb{X}_{N,i} \stackrel{\text{def}}{=} \mathbb{X}_N^{(1)} + \mathbb{X}_{N,i}^{(2)}$. Then the forecasting results will form a sample $\widetilde{x}_{N+M,i}^{(1)}$ $(1 \leq i \leq Q)$, which should be compared against $x_{N+M}^{(1)}$. In this way, *Monte Carlo confidence bounds* for the forecast can be build.

Since in practice we do not know the signal $\mathbb{X}_N^{(1)}$, we cannot apply this procedure. Let us describe the bootstrap procedure for constructing the confidence bounds for the forecast (for a general methodology of bootstrap, see, for example, [10, Sect. 5]).

For a suitable window length L and the grouping of eigentriples, we have the representation $\mathbb{X}_N = \widetilde{\mathbb{X}}_N^{(1)} + \widetilde{\mathbb{X}}_N^{(2)}$, where $\widetilde{\mathbb{X}}_N^{(1)}$ (the reconstructed series) approximates $\mathbb{X}_N^{(1)}$, and $\widetilde{\mathbb{X}}_N^{(2)}$ is the residual series. Suppose now that we have a (stochastic) model of the residuals $\widetilde{\mathbb{X}}_N^{(2)}$. For instance, we can postulate some model for $\mathbb{X}_N^{(2)}$ and, since $\widetilde{\mathbb{X}}_N^{(1)} \approx \mathbb{X}_N^{(1)}$, apply the same model for $\widetilde{\mathbb{X}}_N^{(2)}$ with the estimated parameters. Then, simulating Q independent copies $\widetilde{\mathbb{X}}_{N,i}^{(2)}$ of the series $\mathbb{X}_N^{(2)}$, we obtain Q series $\mathbb{X}_{N,i} \stackrel{\text{def}}{=} \widetilde{\mathbb{X}}_N^{(1)} + \widetilde{\mathbb{X}}_{N,i}^{(2)}$ and produce Q forecasting results $\widetilde{x}_{N+M,i}^{(1)}$ in the same manner as in the straightforward Monte Carlo simulation.

More precisely, any time series $\mathbb{X}_{N,i}$ produces its own reconstructed series $\widetilde{\mathbb{X}}_{N,i}^{(1)}$ and its own forecasting linear recurrence relation LRR_i for the same window length

L and the same set of the eigentriples. Starting at the last $L - 1$ terms of the series $\widetilde{\mathbb{X}}_{N,i}^{(1)}$, we perform M steps of forecasting with the help of its LRR$_i$ to obtain $\widetilde{x}_{N+M,i}^{(1)}$.

As soon as the sample $\widetilde{x}_{N+M,i}^{(1)}$ $(1 \leq i \leq Q)$ of the forecasting results is obtained, we can calculate its (empirical) lower and upper quantiles of some fixed level γ and obtain the corresponding confidence interval for the forecast. This interval will be called the *bootstrap confidence interval*. Simultaneously with the bootstrap confidence intervals for the signal forecasting values, we obtain the bootstrap confidence intervals for the reconstructed values. The average of the bootstrap forecast sample (*bootstrap average forecast*) estimates the mean value of the forecast, while the mean square deviation of the sample shows the accuracy of the estimate.

The simplest model for $\widetilde{\mathbb{X}}_{N}^{(2)}$ is the model of Gaussian white noise. The corresponding hypothesis can be checked with the help of the standard tests for randomness and normality. Another natural approach for noise generation uses the empirical distribution of the residual; this version is implemented in the RSSA package (see [24, Sect. 3.2.1.5]).

3.4.2 Confidence Intervals: Comparison of Forecasting Methods

The aim of this section is to compare different SSA forecasting procedures using several artificial series and the Monte Carlo confidence intervals.

Let $\mathbb{X}_N = \mathbb{X}_N^{(1)} + \mathbb{X}_N^{(2)}$, where $\mathbb{X}_N^{(2)}$ is Gaussian white noise with standard deviation σ. Assume that the signal $\mathbb{X}_N^{(1)}$ admits a recurrent continuation. We shall perform a forecast of the series $\mathbb{X}_N^{(1)}$ for M steps using different versions of SSA forecasting and appropriate eigentriples associated with $\mathbb{X}_N^{(1)}$. Several effects will be illustrated in the proposed simulation study. First, we shall compare some forecasting methods from the viewpoint of their accuracy. Second, we shall demonstrate the role of the proper choice of the window length.

We will consider two examples. In both of them, $N = 100$, $M = 50$ and the standard deviation of the Gaussian white noise $\mathbb{X}_N^{(2)}$ is $\sigma = 1$. The confidence intervals are obtained in terms of the 2.5% upper and lower quantiles of the corresponding empirical c.d.f. using the sample size $Q = 10000$.

Periodic signal: recurrent and vector forecasting. Let us consider a periodic signal $\mathbb{X}_N^{(1)}$ of the form

$$x_n^{(1)} = \sin(2\pi n/17) + 0.5 \sin(2\pi n/10).$$

The series $\mathbb{X}_N^{(1)}$ has difference dimension 4, and we use four leading eigentriples for its forecasting under the choice $L = 50$. The initial series $\mathbb{X}_N = \mathbb{X}_N^{(1)} + \mathbb{X}_N^{(2)}$ and the signal $\mathbb{X}_N^{(1)}$ (red thick line) are depicted in Fig. 3.2a.

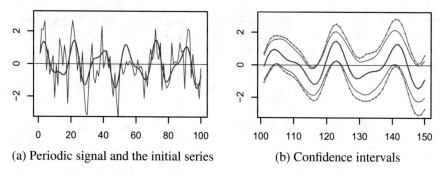

(a) Periodic signal and the initial series (b) Confidence intervals

Fig. 3.2 Comparison of recurrent and vector forecasts

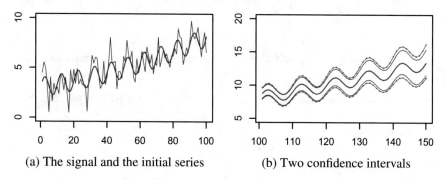

(a) The signal and the initial series (b) Two confidence intervals

Fig. 3.3 Separability and forecasting

Let us apply the Monte Carlo simulation for Basic SSA recurrent and vector forecasting algorithms. Figure 3.2b shows the confidence Monte Carlo intervals for both methods and the true continuation of the signal $\mathbb{X}_N^{(1)}$ (red thick line). Confidence intervals for R-forecasting are marked by dots, while blue thin lines correspond to V-forecasting. We can see that these intervals practically coincide for relatively small numbers of forecasting steps, while V-forecasting has some advantage in the long-term forecasting.

Separability and forecasting. Consider the series $\mathbb{X}_N^{(1)}$ with

$$x_n^{(1)} = 3a^n + \sin(2\pi n/10), \quad a = 1.01.$$

This series is governed by an LRR of dimension 3. Consider Basic SSA R-forecasting for up to 50 points of the signal values $x_{N+j}^{(1)}$ using the series $\mathbb{X}_N = \mathbb{X}_N^{(1)} + \mathbb{X}_N^{(2)}$. We compare two window lengths, $L = 15$ and $L = 50$. The first three eigentriples are chosen for the reconstruction in both choices of L. The series $\mathbb{X}_N$ and the signal $\mathbb{X}_N^{(1)}$ (red thick line) are depicted in Fig. 3.3a.

Figure 3.3b shows that the Monte Carlo forecasting confidence intervals for $L = 15$ (green thin line with dots) are apparently wider than that for $L = 50$. This is not surprising since the choice $L = 50$ corresponds to better separability. This is confirmed by comparing the values of the separability characteristics. In particular, the **w**-correlation (2.18) between the extracted signal and the residual is equal to 0.0083 for $L = 15$ and it equals 0.0016 for $L = 50$. Recall that the exact separability gives zero value for the **w**-correlation.

3.5   Summary and Recommendations on Forecasting Parameters

Let us summarize the material of the previous sections taking as an example Basic SSA R-forecasting method. Other versions of SSA forecasting can be described and commented on similarly.

1. *Statement of the problem*
   We have a time series $\mathbb{X}_N = \mathbb{X}_N^{(1)} + \mathbb{X}_N^{(2)}$ and need to forecast its component $\mathbb{X}_N^{(1)}$.
2. *The main assumptions*

   - The series $\mathbb{X}_N^{(1)}$ admits a recurrent continuation with the help of an LRR of a relatively small dimension r.
   - There exists L such that the series $\mathbb{X}_N^{(1)}$ and $\mathbb{X}_N^{(2)}$ are approximately strongly separable for the window length L.

3. *Proper choice of parameters*
   Since we have to select the window length L providing a sufficient quality of separability and to find the eigentriples corresponding to $\mathbb{X}_N^{(1)}$, all the major rules and recommendations for the use of Basic SSA are applicable here. Note that in this case we must separate $\mathbb{X}_N^{(1)}$ from $\mathbb{X}_N^{(2)}$, but we do not need to obtain a detailed decomposition of the series $\mathbb{X}_N$.
4. *Specifics and potential pitfalls*

   - Since SSA forecasting procedure needs an estimation of the LRR, some recommendations concerning the window length for reconstruction and forecasting can differ. SSA modifications that use different window lengths for reconstruction and for building the forecasting formula can be used.
   - In Basic SSA, if we enlarge the set of proper eigentriples by some extra eigentriples with small singular values, then the result of reconstruction will essentially be the same. When dealing with forecasting, such an operation can produce large perturbations since the trajectory space $\mathcal{X}^{(L,1)}$ will be perturbed a lot; its dimension will be enlarged, and therefore the LRR governing the forecast will be modified. In this case, the magnitude of the extra singular values is not important but the location of the extraneous roots of the characteristic polynomials is important.

5. *Characteristics of forecasting*
   The following characteristics may be helpful in judging the forecasting quality.

- *Separability characteristics.* All separability characteristics considered in Sect. 2.3.3 are of importance for forecasting.
- *Polynomial roots.* The roots of the characteristic polynomial of the forecasting LRR can give an insight into the behaviour of the forecast. These polynomial roots can be useful in answering the following two questions:

  (a) We expect that the forecast has some particular form (for example, we expect it to be increasing). Do the polynomial roots describe such a possibility? For instance, an exponential growth has to be indicated by a single real root (slightly) greater than 1 but if we try to forecast the annual periodicity, then pairs of complex roots with frequencies $\approx k/12$ have to exist.

  (b) Although extraneous roots of the true min-norm LRR have moduli smaller than 1, the extraneous roots of the estimated LRR can be larger than 1. Since the polynomial roots with moduli greater than 1 correspond to the series components with increasing envelopes (see Sect. 3.2), large extraneous roots may cause problems even in the short-term forecasting. This is a serious pitfall that always has to be closely monitored.

- *Verticality coefficient.* The verticality coefficient ν^2 is the squared cosine of the angle between the space $\mathcal{L}_r$ and the vector $\mathbf{e}_L$. The condition $\nu^2 < 1$ is necessary for forecasting. The norm of the min-norm LRR (3.18) coefficients is equal to $\nu^2/(1 - \nu^2)$. This characteristic reflects the ability of the LRR to decrease the noise level, see Proposition 3.4. If ν^2 is close to 1, then the norm is very large. This often means that too many extra eigentriples are taken for the reconstruction of $\mathbb{X}_N^{(1)}$ (alternatively, the whole approach is inadequate).

6. *The role of the initial data*

   Apart from the number M of forecast steps, the formal parameters of Basic SSA R-forecasting algorithm are the window length L and the set I of eigentriples describing $\mathbb{X}_N^{(1)}$. These parameters determine both the forecasting LRR (3.1) and the initial data used in the forecasting formula. Evidently, the forecasting result significantly depends on this data, especially when the forecasting LRR has extraneous roots.

   The SSA R-forecasting method uses the last $L - 1$ terms $\widetilde{x}_{N-L+2}^{(1)}, \ldots, \widetilde{x}_N^{(1)}$ of the reconstructed series $\widetilde{\mathbb{X}}_N^{(1)}$ as the initial data for forecasting. In view of the properties of the diagonal averaging, the last (and the first) terms of the series $\mathbb{X}_N^{(1)}$ are usually reconstructed with poorer precision than the middle ones. This effect may cause substantial forecasting errors. For example, any linear (and nonconstant) series $x_n = an + b$ is governed by the minimal LRR $x_n = 2x_{n-1} - x_{n-2}$, which does not depend on a and b. The parameters a and b used in the forecast are completely determined by the initial data x_1 and x_2. Evidently, errors in this data may considerably modify the forecast.

   Thus, it is important to check the last points of the reconstructed series (for example, to compare them with the expected future behaviour of the series $\mathbb{X}_N^{(1)}$). Even the use of the last points of the initial series as the initial data in the forecasting formula may improve the forecast.

7. *Reconstructed series and LRRs*

In the situation of strong separability between $\mathbb{X}_N^{(1)}$ and $\mathbb{X}_N^{(2)}$ and proper eigen-triple selection, the reconstructed series is governed by the LRR which exactly corresponds to the series $\mathbb{X}_N^{(1)}$. Discrepancies in this correspondence indicate on possible errors: insufficient separability (which can be caused by a bad choice of the forecasting parameters) or general inadequacy of the model. We can suggest the following ways of testing for the presence of these errors and reducing them.

- *Global discrepancies.* Rather than using an LRR for forecasting, we can use it for approximation of either the whole reconstructed series or its subseries. For instance, if we take the first terms of the reconstructed series as the initial data (instead of the last ones) and make $N - L + 1$ steps of the procedure, we can check whether the reconstructed series can be globally approximated with the help of the LRR.
- *Local discrepancies.* The procedure above corresponds to the long-term forecasting. To check the short-term correspondence of the reconstructed series and the forecasting LRR, one can apply a slightly different method which is called the multistart recurrent continuation. In it, for a relatively small M we perform M steps of the multistart recurrent continuation procedure, modifying the initial data from $(\widetilde{x}_1^{(1)}, \ldots, \widetilde{x}_{L-1}^{(1)})$ to $(\widetilde{x}_{K-M+1}^{(1)}, \ldots, \widetilde{x}_{N-M}^{(1)})$, $K = N - L + 1$. The M-step continuation is computed with the help of the forecasting LRR. The results should be compared with $\widetilde{x}_{L+M-1}^{(1)}, \ldots, \widetilde{x}_N^{(1)}$. Since both the LRR and the initial data have errors, the local discrepancies for small M are usually more informative than the global ones. Moreover, by using different M we can estimate the maximal number of steps for a reasonable forecast.

Note that if the discrepancies are small then this does not necessarily imply that the forecasting is accurate. This is because the forecasting LRR is tested on the same points that were used for the calculation of the forecasting LRR.

8. *Forecasting stability and reliability*

While the correctness of the forecast cannot be checked using the data only, the reliability of the forecast can be examined. Let us mention several methods for carrying out such an examination.

- *Different algorithms.* We can try different forecasting algorithms (for example, recurrent and vector) with the same parameters. If their results approximately coincide, we have an argument in favour of forecasting stability.
- *Different window lengths.* If the separability characteristics are stable under small variation in the window length L, we can compare the forecasts for different L.
- *Forecasting of truncated series.* We can truncate the initial series $\mathbb{X}_N$ by removing the last few terms from it. If the separability conditions are stable under this operation, then we can forecast the truncated terms and compare the result with the initial series $\mathbb{X}_N$ and the reconstructed series $\widetilde{\mathbb{X}}_N^{(1)}$ obtained without truncation. If the forecast is regarded as adequate, then its continuation by the same LRR can be regarded as reliable.

9. *Confidence intervals*

Confidence intervals discussed in Sect. 3.4 give important additional information about the accuracy and stability of forecasts.

3.6 Case Study: 'Fortified Wine'

To illustrate the SSA forecasting techniques, we consider the time series 'Fortified wine' (monthly volumes of fortified wine sales in Australia from January 1984 till June 1994, Fig. 2.16). Naturally, time series forecasting should be based on the preliminary time series investigation. We examine both the initial time series of length 174 and its subseries consisting of the first 120 points. We name the former FORT174 and the latter FORT120.

SSA forecasting should only be applied to a time series governed (may be approximately) by some LRR. Therefore, we start with the study of the series from this point of view.

Linear Recurrence Relation Governing the Time Series

Preliminary analysis shows that the 'FORT174' time series (see Sects. 2.3.1.2 and 2.4.2.2) can be decomposed into a sum of a signal and a noise. For window length $L = 84$, the signal can be reconstructed by means of ET1–11 and the **w**-correlation between the signal component and the noise component is 0.004 which is small enough. Thus, the estimated signal subspace of R^L has dimension 11, the min-norm LRR has dimension $L - 1$ and the reconstructed time series (the signal) can be approximated by a time series governed by this LRR. For the series FORT120 and $L = 60$ the signal also corresponds to ET1–11, the **w**-correlation with the residual is slightly larger (it equals 0.005).

Table 3.1 presents the information for 19 leading roots of the characteristic polynomials corresponding to two estimated min-norm LRRs. The roots (recall that they are complex numbers) are ordered by decreasing their moduli. The label 'compl.' for the 'Type' column of Table 3.1 notes that this line relates to two conjugate complex roots $\rho_j e^{\pm i 2\pi \omega_j}$, $0 < \omega_j < 0.5$. In this case, the period $1/\omega_j$ is listed in the table. The first six rows can be interpreted easily: the rows 1–3 and 5–6 correspond to conjugate complex roots, which produce harmonics with periods 6, 4, 2.4, 12, and 3. Moduli larger than one correspond to harmonics with increasing amplitudes, a modulus smaller than one yield a decreasing amplitude. The forth row of the table corresponds to the real-valued root with modulus 0.997. There are no more signal roots and all other roots are extraneous. All moduli of the extraneous roots are less than one. The columns marked 'ET' indicate the correspondence between the eigentriples and the polynomial roots.

The series is decreasing and therefore the roots with modulus larger than 1 are most probably inadequate. Especially, the leading root (ET6–7) has modulus 1.013 for FORT120 which is a possible reason for an unstable forecast. Also, for FORT120

Table 3.1 Time series FORT174 and FORT120: the leading roots of the characteristic polynomials for the min-norm LRRs

FORT174, $L = 84$					FORT120, $L = 60$				
N	ET	Modulus	Period	Type	N	ET	Modulus	Period	Type
1	6–7	1.003	5.969	Compl.	1	6–7	1.013	5.990	Compl.
2	8–9	1.000	3.994	Compl.	2	8–11	1.007	2.376	Compl.
3	4–5	0.998	2.389	Compl.	3	4–5	1.000	4.001	Compl.
4	1	0.997	No	Real	4	1	0.997	No	Real
5	2–3	0.994	12.002	Compl.	5	2–3	0.994	12.033	Compl.
6	10–11	0.989	3.028	Compl.	6	8–11	0.982	3.002	Compl.
7		0.976	3.768	Compl.	7		0.968	5.311	Compl.
8		0.975	3.168	Compl.	8		0.966	9.635	Compl.
9		0.975	10.212	Compl.	9		0.966	3.688	Compl.
10		0.975	5.480	Compl.	10		0.965	2.268	Compl.

two harmonics are mixed; therefore, two pairs of conjugated roots put into correspondence with four eigentriples ET8–11.

Let us check whether the time series FORT174 is well fitted by the estimated min-norm LRR. The maximum value of the global discrepancy between the reconstructed signal and its approximation by a time series governed by the used LRR (that is, the error of global approximation) is equal to 132 and it is smaller than 10% of the time series values. Note that we use the first 83 points as the initial data for the LRR; so the approximation error is calculated starting from the 84-th point.

Let us consider the minimal LRR of order 11 generated by the estimated signal roots presented in Table 3.1 above the horizontal line. (Recall that there is a one-to-one correspondence between LRRs and the roots of the associated characteristic polynomials.) If we take the points 73–83 as the initial data for this LRR, the series governed by the minimal LRR better approximates the time series (the maximum discrepancy is equal to 94). Thus we conclude that the time series is well approximated by the time series governed by the minimal LRR of order 11. Note that since the long-term forecast by the minimal LRR is very sensitive to the initial data, the choice of points 73–83 as the initial data was rather fortunate. The results for local approximation (discrepancy) are similar (magnitudes of errors are smaller while using the minimal LRR).

Since we know the exact period of the time series periodical component (due to its seasonal behavior), we can adjust the LRR by changing the roots so that they correspond to the periods 6, 4, 2.4, 12 and 3. This LRR of order 11 is called an adjusted minimal LRR. The local approximation errors, corresponding to the adjusted minimal LRR, are slightly smaller than for the minimal LRR.

The analytic form of the time series governed by the adjusted minimal LRR is

Table 3.2 Time series FORT120: relative MSD errors of the reconstruction and forecasts

ET	rec (%)	vec12 (%)	rec12 (%)	rec_init12 (%)	vec54 (%)	rec54 (%)	rec_init54 (%)
1	23.11	23.34	23.46	23.49	23.84	23.73	24.02
3	14.79	15.82	16.19	16.41	17.60	17.78	18.17
5	11.63	15.49	15.58	15.44	15.23	15.23	15.57
7	9.70	14.13	15.65	14.41	15.12	24.98	23.26
11	7.45	16.76	17.48	15.59	21.34	23.30	20.57

$$y_n = C_1 0.997^n + C_2 0.994^n \sin(2\pi n/12 + \varphi_2) + $$
$$+ C_3 \sin(2\pi n/4 + \varphi_3) + C_4 1.003^n \sin(2\pi n/6 + \varphi_4) + $$
$$+ C_5 0.998^n \sin(2\pi n/2.4 + \varphi_5) + C_6 0.989^n \sin(2\pi n/3 + \varphi_6).$$

The coefficients C_i and φ_i can be estimated by the least-squares method, see [15, Sect. 3.3] and Fragment 3.5.9 in [24]

The terms are ordered by their eigenvalue shares (in the order of their decreasing). Recall that ordering by roots moduli is generally different from ordering by eigenvalues, since roots moduli are related to the rates of increase/decrease of the time series components and thereby influence a future behavior of the time series governed by the corresponding LRR.

Thus, a preliminary investigation implies that the time series FORT174 and FORT120 well fit to the respective models of the form required, so we can start their forecasting.

Choice of Forecasting Methods and Parameters

Let us demonstrate the approach to forecasting on the 'Fortified wine' example, investigating the accuracy of forecasting the values at the points 121–174 (the test period) on the base of the reconstruction of the points 1–120 (the base period 'FORT120'). The 12-point ahead and 54-point ahead forecasts are considered. Table 3.2 summarizes the errors of forecasts for different forecasting methods. The relative MSD errors of estimation of $\mathbb{Y}$ by $\widetilde{\mathbb{Y}}$ are calculated as

$$\|\widetilde{\mathbb{Y}} - \mathbb{Y}\|_F / \|\mathbb{Y}\|_F \cdot 100\%. \tag{3.19}$$

In Table 3.2, the column 'ET' shows the chosen numbers of the leading eigentriples, the column 'rec' gives the reconstruction errors, the columns 'vec12', 'rec12', 'vec54', 'rec54' correspond to vector and recurrent forecasting for the horizons 12 and 54 terms respectively. The suffix '_init' means that the forecasting formula was applied to the initial series rather than to the reconstructed one.

Let us now discuss the main points of the forecasting logic.

1. Note first that only a set of components separated from the residual may be chosen. For the 'FORT120' series the admissible numbers of components are 1, 3, 5, 7, or 11.
2. There is a conflict between the accuracy of reconstruction and stability of forecasting. In Table 3.2 the errors of reconstruction decrease (the column 'rec') while the errors of forecasts decrease in the beginning and increase later. Note that all considered components are related to the signal and therefore the increase of errors is related to instability.
3. The observed behaviour of the forecasting errors means that the optimal number of the components for forecasting is 7 for 12-term ahead and 5 for 54-term ahead forecasts. This is a natural result since the stability of forecasting is much more important for the long-term forecasting.
4. The vector forecasting method provides more stable forecast of 'FORT120' which becomes apparent for long horizons.
5. Comparison of the forecasting methods can be performed by means of the confidence intervals: smaller size of the confidence intervals indicates better stability of forecasting. This approach does not help for choosing the optimal number of components, since the rough forecast can be the most stable. However, this is a good tool to compare forecasting modifications for a fixed number of components. In particular, the size of the bootstrap confidence interval for ET1–7 is one and half times smaller for the vector forecast than that for the recurrent forecast.
6. Table 3.2 shows that generally the forecasting formula can be applied to the initial time series (the columns with the suffix '_init') instead of the reconstructed one. However, there is no noticeable improvement.
7. The LRR can be adjusted by two ways. The first modification is to remove extraneous roots and to adjust the signal roots using prior information. For the 'FORT120' time series we know the periods of the seasonal component. The modified LRR looks more appropriate as it adequately incorporates the periodicity. However, the forecast is very unstable and gives the forecasting error several times larger than for the min-norm LRR. The reason is that the initial data with error can significantly change the amplitudes of the true harmonics. Certainly, the minimal LRR should be applied to the reconstructed series.
8. A specific feature of this dataset is that the behaviour of the series is close to multiplicative. However, this time series is not purely multiplicative since the form of the seasonal period differ from year to year (Fig. 2.17). The last conclusion is confirmed by different moduli of the roots. For the initial time series the leading harmonic with period 12 is decreasing and the estimated modulus of the corresponding root is equal to 0.994. Therefore, the decreasing exponential has stable behaviour, regardless of the estimation errors. After the log-transformation of a multiplicative series, the root modulus becomes close to 1 and the estimation error can give the modulus of the estimated root larger than 1; that is, the forecast (especially, long-term) could be unstable. The 'FORT120' series demonstrates this effect, since the forecasting error for the log-transformed data is larger than that for the original data.

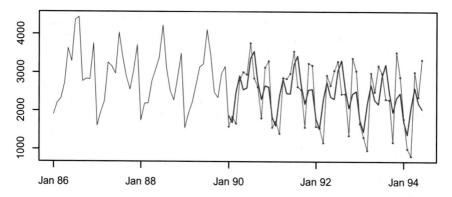

Fig. 3.4 FORT120: two forecasts

Figure 3.4 shows the last 4 years of the series 'FORT120' (blue line) and two vector forecasts for 54 points ahead: the stable and accurate forecast based on ET1–5 (red thick line) and the forecast with unstable and less accurate behaviour based on ET1–11 (green line with dots). Note that the accuracy of forecasting for 12-points ahead is approximately the same for both forecasts.

Summarizing we make the following conclusions concerning our experience with forecasting the 'Fortified wine' series: (1) it is better to use the original time series rather than its log-transformed version; (2) the best eigentriple group used for forecasting is either ET1–5 for long-term forecast or ET1–7 for short-term forecast; (3) V-forecasting is more accurate that R-forecasting.

The conclusion about the accuracy of the forecasts is made on the base of comparison of the forecasts with series values in the forecasted points. Stability of the forecasts can be checked by means of the confidence intervals and does not require the knowledge of the series values. For two considered forecasts, the size of confidence intervals for ET1–11 is more than twice larger than that for ET1–5, if we take the last forecasted year. Thus, this example demonstrates that the more accurate long-term forecast corresponds to the more stable one.

3.7 Imputation of Missing Values

This section is devoted to the extension of SSA forecasting algorithms for the analysis of time series with missing data.

The following three approaches for solving this problem are known. The first approach was suggested in [42]. This approach is suitable for stationary time series only and uses the following simple idea: in the process of the calculation of inner products of vectors with missing components, we use only pairs of valid vector components and omit the others.

The second '*Iterative*' approach uses an iterative interpolation. Initially, the places of missing values are filled with arbitrary numbers. Then these numbers are iteratively refined by the successive application of SSA. After each iteration, the values at the places of missing values are taken from the previous iteration but the other values are taken from the initial time series. This approach can be formally applied for almost any location of missing values. Therefore, several artificial gaps can be added and then be used to justify the choice of SSA parameters, namely, the window length and the number of chosen components. This idea was suggested in [4] for the imputation of missing values in matrices and then was extended to time series in [30]. The iterative approach has a semi-empirical reasoning for convergence. However, even for noiseless signals the gaps cannot be filled in one iteration. Therefore, this method has large computational cost. Also, it does not provide exact imputation and it needs an additional information about the subspace dimension.

The third approach of filling in missing data is an extension of SSA forecasting algorithms. This approach is considered below in this section and is called '*the subspace approach*'. According to this approach we continue the structure of the extracted component to the gaps caused by the missing data [16]. The theory of SSA assumes that the forecasted component is (or is approximated by) a time series of finite rank. Theoretical results concerning the exact reconstruction of missing values are also based on this assumption. Nevertheless, the constructed algorithms are applicable to real-life time series with missing values where they give approximate results.

Note that in a particular case, when the missing values are located at the end of the series, the problem of their filling in coincides with the problem of forecasting.

The subspace-based methods of gap filling are partly implemented in the RSSA package, see [24, Sect. 3.3.3], along with the iterative filling-in algorithm.

The Layout of the Algorithm

The general structure of the algorithm for the analysis of time series with missing data is the same as for Basic SSA, but the steps are somewhat different.

Assume that we have the initial time series $\mathbb{X}_N = (x_1, \ldots, x_N)$ consisting of N elements, with a part of $\mathbb{X}_N$ unknown. Let us describe the algorithm, using the notation of Sect. 3.1.4, in the case of reconstruction of the first component $\mathbb{X}_N^{(1)}$ of the observed series $\mathbb{X}_N = \mathbb{X}_N^{(1)} + \mathbb{X}_N^{(2)}$.

First Stage: Decomposition

Step 1. Embedding. Let us fix the window length L, $1 < L < N$. The embedding procedure transforms the initial time series into the sequence of L-dimensional lagged vectors $\{X_i\}_{i=1}^K$, where $K = N - L + 1$. Some of the lagged vectors may be incomplete, i.e., contain missing components. Let $\mathcal{C}$ be the set of indices such that the lagged vectors X_i with $i \in \mathcal{C}$ are complete. Let us collect all complete lagged vectors X_i, $i \in \mathcal{C}$, into the matrix $\widetilde{\mathbf{X}}$. Assume that this matrix is non-empty. If there are no missing values, then the matrix $\widetilde{\mathbf{X}}$ coincides with the trajectory matrix of the series $\mathbb{X}_N$. Note that the construction of $\widetilde{\mathbf{X}}$ is the same as in Shaped SSA, see Sect. 2.6.3 and [24, Sect. 2.6].

Step 2. Finding the basis. Let $\widetilde{\mathbf{S}} = \widetilde{\mathbf{X}}\widetilde{\mathbf{X}}^{\mathrm{T}}$. Denote by $\lambda_1 \geq \ldots \geq \lambda_L \geq 0$ the ordered eigenvalues of the matrix $\widetilde{\mathbf{S}}$ and by $U_1, \ldots, U_L$ the orthonormal system of the eigenvectors of the matrix $\widetilde{\mathbf{S}}$ corresponding to these eigenvalues, $d = \max\{i : \lambda_i > 0\}$.

Second Stage: Reconstruction

Step 3a. Choosing the subspace and projection of the complete lagged vectors. Let a set of indices $I_r = \{i_1, \ldots, i_r\} \subset \{1, \ldots, d\}$ be chosen and the subspace $\mathcal{M}_r = \mathrm{span}(U_{i_1}, \ldots, U_{i_r})$ be formed. The choice of the eigenvectors (i.e., their indices) corresponding to $\mathbb{X}_N^{(1)}$ is the same as in Basic SSA. The complete lagged vectors can be projected onto the subspace $\mathcal{M}_r$ in the usual way:

$$\widehat{X}_i = \sum_{k \in I_r} (X_i, U_k) U_k, \quad i \in \mathcal{C}.$$

Step 3b. Projection of the incomplete lagged vectors. For each $\mathcal{Q}$-incomplete lagged vector with missing components in the positions from the set $\mathcal{Q}$, we perform this step which consists of two parts:

(α) calculation of $\widehat{X}_i\big|_{J_L \setminus \mathcal{Q}}, \quad i \notin \mathcal{C}$,

(β) calculation of $\widehat{X}_i\big|_{\mathcal{Q}}, \quad i \notin \mathcal{C}$.

Since adjacent lagged vectors have common information (the trajectory matrix (2.1) consisting of the lagged vectors is Hankel) there are many possible ways of solving the formulated problems. Some of these ways will be discussed in the following sections. The available information also enables processing of 'empty' vectors with $\mathcal{Q} = J_L = \{1, \ldots, L\}$. Note that step 3b may change the vectors $\widehat{X}_i, i \in \mathcal{C}$. The result of steps 3a and 3b is the matrix $\widehat{\mathbf{X}} = [\widehat{X}_1 : \ldots : \widehat{X}_K]$, which serves as an approximation to the trajectory matrix of the series $\mathbb{X}_N^{(1)}$, under the proper choice of the set I_r.

Step 4. Diagonal averaging. The matrix $\widehat{\mathbf{X}}$ is transformed into the new series $\widetilde{\mathbb{X}}_N^{(1)}$ (the reconstructed time series) by means of the diagonal averaging.

Clusters of Missing Data

Implementation of step 3b for projecting the incomplete vectors needs a definition of clusters of missing data and their classification assuming that L is fixed.

A sequence of missing data of a time series is called a *cluster of missing data* if every two adjacent missing values from this sequence are separated by less than L non-missing values and there is no missing data among L neighbours (if they exist) on the left/right element of the cluster. Thus, a group of at least L successive non-missing values of the series separates clusters of missing data. A cluster is called *left/right* if its left/right element is located at a distance of less than L from the left/right end of the series. A cluster is called *continuous* if it consists of successive missing data.

Step 3b can be performed independently for each cluster of missing data.

Methods for Step 3b

Different realizations of Step 3b are thoroughly considered in [16]. Here we briefly describe several typical versions and their relation to SSA forecasting methods formulated in Sect. 3.1. Let the window length L and the indices of the eigentriples corresponding to the chosen time series component be fixed. Propositions 3.2 and 3.3 (where we take $n = L$, $m = r$, $I_r = \{i_1, \ldots, i_r\}$, $\mathbf{P} = [U_{i_1} : \ldots : U_{i_r}]$) provide the theoretical ground for the methods of filling in.

If the considered cluster is continuous and is not left, then (3.9) with $\mathcal{Q} = \{L\}$ provides the coefficients of an LRR that can be applied to the reconstructed points that lie on the left from the missing data cluster (*sequential filling in from the left*). Similarly, setting $\mathcal{Q} = \{1\}$ and applying the backward recurrence relation (3.9) to the reconstructed data taken from the right side, *sequential filling in from the right* can be introduced. Different combinations of the sequential fillings in from the left and from the right (the so-called two-sided methods) can be constructed. For example, their average can be used.

Remark 3.7 Consider a continuous cluster of missing data of length M, which is a right extreme cluster (and assume that there are no other clusters of missing data in the series). If the sequential method described above is applied to this cluster, then the result will coincide with the recurrent forecast for M terms ahead (Sect. 3.1.2.1), where the forecast is constructed on the first $N - M$ points of the time series and the same parameters L and I_r.

In the same manner as we have used for the vector forecasting (Sect. 3.1.2.2), the vector coordinates at the positions of non-missing components can be filled with the help of the adjacent complete vectors and then projected to $\mathcal{M}|_{J_L \backslash \mathcal{Q}}$ by the projector given by formula (3.11) ('*$\boldsymbol{\Pi}$ Projector*').

Also, in the same manner as the simultaneous forecasting was introduced (see Sect. 3.1.2.3), the vector coordinates at the positions of missing components can be filled in simultaneously, not one by one as in the sequential filling in, since Proposition 3.2 allows filling in several vector coordinates at once ('*simultaneous filling in*'). This may simplify the imputation of not-continuous clusters of missing data.

Discussion

- As well as for forecasting, the approach above allows filling in missing values in any component of the time series, not necessary in the signal. For example, missing values in the trend can be filled in. Certainly, an approximate separability of the imputed component from the residual is required.
- If the time series component is exactly separated from the residual and has finite rank, it can be filled in exactly.
- The location of missing data is very important for the possibility of imputation by the subspace method, since the number of non-missing values should be large enough for achieving separability. At least, the number of the complete lagged vectors should be larger than the rank of the forecasted time series component.
- If there are many randomly located missing data, then it can be impossible to get a sufficient number of lagged vectors. However, it is possible to estimate the subspace by involving the lagged vectors with a few missing entries; see [16] for details.

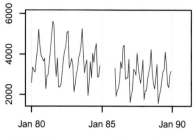   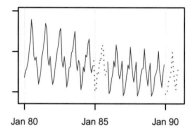

(a) Initial time series with missing data   (b) Reconstructed time series with filled in data

Fig. 3.5 FORT120: Filling in missing data

Example

To demonstrate the work of the methods of filling in missing data, let us consider the time series FORT120, which was investigated for forecasting in Sect. 3.6.

Let us remove 12 known values, starting with 60th point (i.e., we assume that the values for a year since December 1984 are unknown). For this artificially missing data we estimate the accuracy of their recovery for different versions of the algorithm. Also, to simulate forecast, we add 12 missing data after the last, 120th point of the series. The time series obtained is illustrated in Fig. 3.5a.

The first question is how to choose the window length L. In the case of no missing data, the general recommendation is to choose the window length close to $N/2$ and divisible by the period of expected periodicity (here, it is 12 months). The window length $L = 60$ meets these conditions. However, for $L = 60$ all lagged vectors will contain missing data. Hence we have to choose smaller L. The choice of $L = 36$ provides us with 38 complete lagged vectors with no missing data.

The analysis of the time series FORT120 in Sect. 3.6 shows that the eigenvectors with indices 1–7 provide the best forecast for 12 points ahead for the choice $L = 60$, while the whole signal is described by the 11 leading eigentriples. The structure of the eigentriples for $L = 36$ is similar and we can use the interpretation of the leading eigentriples found in Sect. 3.6.

The comparison of the filling in results with the values, which were artificially removed from the initial time series, shows an advantage of the version using 'Π Projector' with simultaneous filling in of the missing data and the choice $r = 11$, $I_r = \{1, 2, \ldots, 11\}$. This differs from the forecasting results obtained in Sect. 3.6. Note that the used method of filling in missing data at the end of the time series was not considered during forecasting. Therefore, the ideas of missing data imputation can extend the number of forecasting methods. However, since the precision of reconstruction influences forecasting accuracy much more than the quality of missing data imputation, when the missing data is somewhere in the middle of the series, some methods developed for missing value imputation can be inappropriate as methods of forecasting; an obvious example is the iterative approach.

The result of missing data imputation is illustrated in Fig. 3.5b. The reconstructed series is marked by the dotted line in the area of missing data. The relative MSD

Table 3.3 FORT120: MSD errors for Iterative and Subspace methods of filling in

Method	Middle	End	Total
Subspace $L = 36$	255.9	292.8	275.0
Iterative $L = 36$	221.2	333.0	282.7
Iterative $L = 60$	216.2	419.3	333.6

error (3.19) of reconstruction is approximately equal to 9% for the missing data and to 6% for the non-missing terms in the series.

Comparison with the iterative method. Let us apply the iterative method to the same FORT120 data with the same missing entries, at the middle of the series and at the end. If we replace the missing data by the average value of all valid series points, then 20 iterations are sufficient for convergence. The results are presented in Table 3.3. The errors are calculated as square root of the average squared deviations. For the missing values in the middle, the iterative method provides slightly smaller errors of reconstruction than the subspace method, while for the end points (that is, for forecasting) the iterative method is not stable with respect to the window length. Note that the choice $L = 60$ is not appropriate for the subspace method.

Simulations performed for noisy model series of finite rank in the form of a sum of several products of exponential and harmonic series confirm that the error of filling in missing data at the middle are similar for both methods, while the subspace method is more stable for forecasting.

3.8   Subspace-Based Methods and Estimation of Signal Parameters

While the problems of reconstruction and forecasting are traditionally included into the scope of problems solved by SSA, estimation of signal parameters is usually not. In contrast, estimation of signal parameters is the primary objective for many subspace-based methods of signal processing. In this section we follow [14] to describe the most common subspace-based methods and demonstrate their compatibility with SSA. For simplicity of notation we always assume $L \leq K = N - L + 1$.

Let us shortly describe the problem. Consider a signal $\mathbb{S}_N = (s_1, \ldots, s_N)$ in the form $s_n = \sum_{j=1}^{r} c_j \mu_j^n, n = 1, \ldots, N$, where all μ_j are assumed to be different; the more complicated form (3.13) can be considered in a similar manner. The problem is to estimate μ_j observing the noisy signal. The $\mu_j = \rho_j e^{i2\pi\omega_j}$ are expressed in terms of parameters ρ_j and ω_j, which can often be interpreted. In particular, ω_j are the frequencies presented in the signal. An estimator of μ_j provides the information about the structure of the signal, which is different from the information we get from the coefficients c_j. Note that if the time series is real-valued, then s_n can be written as the sum of modulated sinusoids $A_j \rho_j^n \cos(2\pi\omega_j n + \varphi_j)$.

The idea of subspace-based methods is as follows. Let $r < N/2$. The signal $\mathbb{S}_N$ with $s_n = \sum_{j=1}^{r} c_j \mu_j^n$ has rank r and is governed by LRRs like $s_n = \sum_{k=1}^{t} a_k s_{n-k}$, $t \geq r$. Then μ_j can be found as the signal roots of the characteristic polynomial of a governing LRR (see Sect. 3.2). Simultaneously, the L-trajectory space $(L > r)$ of the signal (the so-called signal subspace) has dimension r and is spanned by the vectors $(1, \mu_j, \ldots, \mu_j^{L-1})^{\mathsf{T}}$. The coefficients of the governing LRRs of order $L - 1$ can also be found using the information about the signal subspace. Methods of estimating μ_j based on estimation of the signal subspace are called *subspace-based methods*.

Since finding signal roots of the characteristic polynomial of the LRR governing the signal is very important for estimation of the signal parameters, we start with several facts relating signal roots to eigenvalues of some matrix.

3.8.1 Basic Facts

The next statement follows from the properties of eigenvalues.

Proposition 3.5 *Roots of a polynomial* $p(\mu) = \mu^M + c_1 \mu^{M-1} + \ldots + c_{M-1}\mu + c_M$ *coincide with eigenvalues of its companion matrix* $\mathbf{C}$ *defined by*

$$\mathbf{C} = \begin{pmatrix} 0 & 0 & \ldots & 0 & -c_M \\ 1 & 0 & \ldots & 0 & -c_{M-1} \\ 0 & 1 & \ldots & 0 & -c_{M-2} \\ \vdots & \vdots & \ddots & \vdots & \vdots \\ 0 & 0 & \ldots & 1 & -c_1 \end{pmatrix}.$$

Note that the multiplicities of the roots of the polynomial $p(\mu)$ are equal to the algebraic multiplicities of the eigenvalues of its companion matrix (i.e., to the multiplicities of the roots of the characteristic polynomial of this matrix). However, these multiplicities do not always coincide with the geometric multiplicities, which are equal to the dimensions of the eigenspaces corresponding to the eigenvalues.

To derive an analytic form of the signal $(s_n = \sum_{j=1}^{t} c_j \mu_j^n$ or (3.13) in the general case), we need to find roots of the characteristic polynomial of the LRR governing the signal. By Proposition 3.5, we have to find either the roots of the characteristic polynomial or the eigenvalues of its companion matrix. The latter does not require the full knowledge of the LRR. Let us demonstrate that for finding the signal roots it is sufficient to know the basis of the signal trajectory space.

Let $\mathbf{C}$ be a full-rank $d \times d$ matrix, $Z \in \mathsf{R}^d$, and $\mathbf{Z}$ be a full-rank $L \times d$ matrix $(L > d)$, which can be expressed as

$$\mathbf{Z} = \begin{pmatrix} \mathbf{Z}^{\mathrm{T}} \\ \mathbf{Z}^{\mathrm{T}}\mathbf{C} \\ \vdots \\ \mathbf{Z}^{\mathrm{T}}\mathbf{C}^{L-1} \end{pmatrix}. \tag{3.20}$$

Let us again denote the matrix $\mathbf{Z}$ without the last row by $\underline{\mathbf{Z}}$ and the matrix $\mathbf{Z}$ without its first row by $\overline{\mathbf{Z}}$. It is clear that $\overline{\mathbf{Z}} = \underline{\mathbf{Z}}\mathbf{C}$. We call this property of $\mathbf{Z}$ the *shift property* generated by the matrix $\mathbf{C}$.

Proposition 3.6 *Let* $\mathbf{Z}$ *satisfy the shift property generated by the matrix* $\mathbf{C}$, $\mathbf{P}$ *be a full-rank* $d \times d$ *matrix, and* $\mathbf{Y} = \mathbf{ZP}$. *Then the matrix* $\mathbf{Y}$ *satisfies the shift property generated by the matrix* $\mathbf{D} = \mathbf{P}^{-1}\mathbf{CP}$, *i.e.,* $\overline{\mathbf{Y}} = \underline{\mathbf{Y}}\mathbf{D}$.

The proof of this proposition is straightforward.

Note that the multiplication by a nonsingular matrix $\mathbf{P}$ can be considered as a transformation of the vector coordinates in the column space of the matrix $\mathbf{Z}$. It is easily seen that the matrices $\mathbf{C}$ and $\mathbf{D} = \mathbf{P}^{-1}\mathbf{CP}$ have the same eigenvalues; these matrices are called *similar*.

Remark 3.8 Let the matrix $\mathbf{Y}$ satisfy the shift property generated by the matrix $\mathbf{D}$. Then $\mathbf{D} = \underline{\mathbf{Y}}^{\dagger}\overline{\mathbf{Y}}$, where $\mathbf{A}^{\dagger}$ denotes the Moore–Penrose pseudoinverse of $\mathbf{A}$.

Proposition 3.7 *Let a time series* $\mathbb{S}_N = (s_1, \ldots, s_N)$ *satisfy the minimal LRR (3.12) of order* d, $L > d$ *be the window length,* $\mathbf{C}$ *be the companion matrix of the characteristic polynomial of this LRR. Then any* $L \times d$ *matrix* $\mathbf{Y}$ *with columns forming a basis of the trajectory space of* $\mathbb{S}_N$ *satisfies the shift property generated by some matrix* $\mathbf{D}$. *Moreover, the eigenvalues of this shift matrix* $\mathbf{D}$ *coincide with the eigenvalues of the companion matrix* $\mathbf{C}$ *and hence with the roots of the characteristic polynomial of the LRR.*

Proof Note that for any $1 \le i \le N - d$ we have

$$(s_i, s_{i+1}, \ldots, s_{i+(d-1)})\mathbf{C} = (s_{i+1}, s_{i+2}, \ldots, s_{i+d}).$$

Therefore, (3.20) holds for $\mathbf{Z} = (x_1, x_2, \ldots, s_d)^{\mathrm{T}}$. It can be easily proved that for a time series governed by the minimal LRR of order d, any d adjacent columns of the trajectory matrix are linearly independent. Consequently, the matrix $\mathbf{Z} = [S_1 : \ldots : S_d]$ is of full rank and we can apply Proposition 3.6. $\qquad\square$

Remark 3.9 The SVD of the L-trajectory matrix of a time series provides a basis of its trajectory space. Specifically, the left singular vectors which correspond to the nonzero singular values form such a basis. If we observe a time series of the form 'signal + residual', then the SVD of its L-trajectory matrix provides the basis of the signal subspace under the condition of exact strong separability of the signal and the residual.

3.8.2 ESPRIT

Consider a time series $\mathbb{X}_N = \{x_i\}_{i=1}^N$ with $x_i = s_i + p_i$, where $\mathbb{S}_N = \{s_i\}_{i=1}^N$ is a time series governed by an LRR of order r (that is, signal) and $\mathbb{P}_N = \{p_i\}_{i=1}^N$ is a residual (noise, perturbation). Let $\mathbf{X}$ be the trajectory matrix of $\mathbb{X}_N$. In the case of exact or approximate separability of the signal and the residual, there is a set I of eigenvector indices in (2.2), which correspond to the signal. If the signal dominates, then $I = \{1, \ldots, r\}$ and the subspace $\mathcal{L}_r = \text{span}\{U_1, \ldots, U_r\}$ can be considered as an estimate of the true signal subspace $\mathbb{S}$. Therefore, we can use $\widetilde{\mathbf{Y}} = \mathbf{U}_r = [U_1 : \ldots : U_r]$ as an estimate of $\mathbf{Y}$ from Proposition 3.7. Then the shift property is approximately met and $\underline{\mathbf{U}}_r \mathbf{D} \approx \overline{\mathbf{U}}_r$.

The method ESPRIT consists in estimation of the signal roots as the eigenvalues of a matrix $\widehat{\mathbf{D}}$, for which

$$\underline{\mathbf{U}}_r \widehat{\mathbf{D}} \approx \overline{\mathbf{U}}_r. \tag{3.21}$$

By estimating the signal roots, ESPRIT provides estimates of the signal parameters. See how to call the corresponding R functions from the RSSA package in [24, Sect. 3.1.3]

Let us study the methods of finding the matrix $\widehat{\mathbf{D}}$. The main idea of LS-ESPRIT was introduced in the paper [33] devoted to the problem of estimating frequencies in a sum of sinusoids, in the presence of noise. The method was given the name ESPRIT in [40]; this name was later used in many other papers devoted to the DOA (Direction of Arrival) problem. For time series processing, LS-ESPRIT is also called Hankel SVD (HSVD, [3]). Later the so-called TLS-ESPRIT modification was suggested (see e.g. [48], where the method was called Hankel Total Least Squares (HTLS)). There are papers devoted to the perturbation study of ESPRIT, see e.g. [2], where specific features of ESPRIT in the case of multiple roots are also described.

Remark 3.10 ESPRIT is able to estimate parameters of a separable time series component, not necessary the signal, if the matrix $\mathbf{U}_r$ consists of the corresponding eigenvectors.

Least Squares (LS-ESPRIT). The LS-ESPRIT estimate of the matrix $\mathbf{D}$ is

$$\widehat{\mathbf{D}} = \underline{\mathbf{U}}_r^\dagger \overline{\mathbf{U}}_r = (\underline{\mathbf{U}}_r^{\mathrm{T}} \underline{\mathbf{U}}_r)^{-1} \underline{\mathbf{U}}_r^{\mathrm{T}} \overline{\mathbf{U}}_r. \tag{3.22}$$

The eigenvalues of $\widehat{\mathbf{D}}$ do not depend on the choice of the basis of the subspace $\mathcal{L}_r = \text{span}\{U_1, \ldots, U_r\}$.

Total Least Squares (TLS-ESPRIT). As $\mathbf{U}_r$ is known only approximately then there are errors in both $\underline{\mathbf{U}}_r$ and $\overline{\mathbf{U}}_r$. Therefore, the solution of the approximate equality $\underline{\mathbf{U}}_r \mathbf{D} \approx \overline{\mathbf{U}}_r$, based on the method of Total Least Squares (TLS), can be more accurate.

Recall that to solve the equation $\mathbf{AX} \approx \mathbf{B}$, TLS minimizes the sum

$$\|\widetilde{\mathbf{A}} - \mathbf{A}\|_{\mathrm{F}}^2 + \|\widetilde{\mathbf{B}} - \mathbf{B}\|_{\mathrm{F}}^2 \longrightarrow \min \tag{3.23}$$

with respect to $\widetilde{\mathbf{A}}$ and $\widetilde{\mathbf{B}}$ such that $\exists \mathbf{Z} : \widetilde{\mathbf{A}}\mathbf{Z} = \widetilde{\mathbf{B}}$.

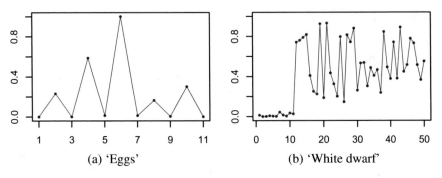

(a) 'Eggs'   (b) 'White dwarf'

Fig. 3.6 Rank and separability detection by ESTER

Set $\mathbf{A} = \underline{\mathbf{U}}_r$, $\mathbf{B} = \overline{\mathbf{U}}_r$ in (3.23). Then the matrix $\mathbf{Z}$ that minimizes (3.23) is called the TLS-estimate of $\mathbf{D}$ (see [26] for explicit formulas). The TLS-ESPRIT estimate is the same for any orthonormal basis in $\mathcal{L}_r$, see [14]. This is not true if the basis in $\mathcal{L}_r$ is not orthogonal.

ESPRIT and rank estimation. ESPRIT deals with the matrix equation (3.21), which has a solution if the r leading components are exactly separated from the residual (the remaining $L - r$ components). Therefore, some measure of difference between the left-hand and the right-hand sides of the matrix equation can indicate the cut-off points of separability; that is, it can suggest the number of the leading SVD components that are separated from the residual. Therefore, the last cut-off point of separability corresponds to the rank estimation. In [1], the L_2-norm of the difference, we denote it $\rho_2(r)$, is used for estimating the rank of the signal (the ESTER method) provided that there are no separability points within the signal components. An attractive feature of the ESTER-type methods is that they assume only separability of the signal from noise and do not assume parametric forms for the signal and noise.

However, the ESTER-type estimates of rank appear to be unstable for real-world time series. Figure 3.6a shows how the ESTER reflects the points of separability: small values of $\rho_2(r)$ correspond to the cut-off points of separability (compare with Fig. 2.24a). In Fig. 3.6b, where the rank of the signal is estimated to be 11 (see Fig. 2.26), the behavior of $\rho_2(r)$ demonstrates that there are no small values of $\rho_2(r)$ for $r \leq 11$.

3.8.3 Overview of Other Subspace-Based Methods

In this subsection, we demonstrate ideas of other subspace-based methods which are different from SSA and ESPRIT-like methods. These methods are applied to time series governed by LRRs and estimate the main (signal) roots of the corresponding characteristic polynomials. The most fundamental subspace-based methods were developed for the cases of a noisy sum of imaginary exponentials (cisoids) and of real sinusoids, for the purpose of the estimating their frequencies, see e.g. [45]. We are mostly interested in the methods that can be applied to any time series of finite rank given in the form (3.13). Below we write down familiar methods in a unified notation; references can be found in [45].

We start with a description of general methods in the complex-valued case. Most of these general methods use the correspondence between LRRs that govern the signal and vectors from the subspace orthogonal to the signal subspace, which was introduced in Sect. 3.2.2.

The first idea is to use the properties of signal and extraneous roots to distinguish between them. Let us introduce three possible realizations of this idea. As before, $\mathcal{S}$ is the signal subspace and $\mathcal{L}_r$ is its estimate.

Version 1. Consider an LRR that governs the signal (the best choice is the min-norm LRR, see Sect. 3.2.2; however, this is not essential). Then find all the roots μ_m of the characteristic polynomial of this LRR and then find coefficients c_{mj} in (3.13). The coefficients c_{mj} corresponding to the extraneous roots are equal to 0. In the case of a noisy signal, $\widehat{\mu}_m$ are the roots of a polynomial with coefficients taken from a vector that belongs to $\mathcal{L}_r^\perp$, and the extraneous roots have small absolute values of the LS estimates $\widehat{c}_{mj}$.

Version 2. Let us consider the forward and backward min-norm predictions. It is known that the corresponding characteristic polynomials have the conjugate extraneous roots and their signal roots are reciprocal, that is, connected by the relation $z' = z^{-1}$, see [5, Proposition 2]. Note that the forward prediction given by a vector $A \in \mathcal{S}^\perp$ corresponds to the roots of $\langle Z(z), A \rangle = 0$, where $Z(z) = (1, z, \ldots, z^{L-1})^\mathsf{T}$ and $\langle \cdot, \cdot \rangle$ is the inner product in the complex Euclidean space. At the same time, the backward prediction given by a vector $B \in \mathcal{S}^\perp$ corresponds to the roots of $\langle Z(1/z), B \rangle = 0$. If we consider the roots of the forward and backward min-norm polynomials together, then all extraneous roots lie inside the unit circle, while one of z' and z is located on or outside the unit circle. This allows us to detect the signal roots. For a noisy signal, A and B are specific vectors taken from $\mathcal{L}_r^\perp$: these vectors are the projections onto $\mathcal{L}_r^\perp$ of the unit vectors $\mathbf{e}_L$ and $\mathbf{e}_1$, correspondingly. If all coefficients of a polynomial are real, then the set of roots coincides with the set of their complex conjugates due to properties of roots of polynomials with real coefficients. However, if complex-valued time series and complex coefficients are considered, then it is convenient to consider the backward min-norm polynomials with conjugate coefficients; the corresponding LRR governs the reverse series of complex conjugates to the original time series.

Version 3. Let us take a set of vectors from $\mathcal{S}^\perp$. Each vector from $\mathcal{S}^\perp$ with nonzero last coordinate generates an LRR. The signal roots of the characteristic polynomials of these LRRs are equal, whereas the extraneous roots are arbitrary. For a noisy signal, the set of vectors is taken from $\mathcal{L}_r^\perp$. Then the signal roots correspond to clusters of roots if we consider pooled roots.

Several more methods are developed for estimation of frequencies in a noisy sum of undamped sinusoids or imaginary exponentials. Let for simplicity $s_n = \sum_{k=1}^r c_k e^{\mathrm{i}2\pi\omega_k n}$. In this case, the signal roots $e^{\mathrm{i}2\pi\omega_k}$ all have the absolute value 1 and can be parameterized by one parameter (frequency) only. Let $W = W(\omega) =$

$Z(e^{i2\pi\omega})$. As $W(\omega_k) \in \mathcal{S}$, $\langle W(\omega_k), A \rangle = 0$ for all $A \in \mathcal{S}^\perp$. Therefore, if $A \in \mathcal{S}^\perp$, then we can consider the square of the cosine of the angle between $W(\omega)$ and A as a measure of their orthogonality. This idea forms the basis for the Min-Norm and MUSIC methods. The modifications of the methods in which the roots are ordered by the absolute value of the deviation of their moduli from the unit circle have names with the prefix 'root-'.

Version 4. Min-Norm. Let $f(\omega) = \cos^2(\widehat{W(\omega), A})$, where A, the projection of $\mathbf{e}_L$ onto $\mathcal{L}_r^\perp$, is the vector corresponding to the min-norm forward prediction. The Min-Norm method is based on searching for the maximums of $1/f(\omega)$; this function can be interpreted as a pseudospectrum.

Version 5. MUSIC. Let $f(\omega) = \cos^2(\widehat{W(\omega), \mathcal{L}_r^\perp})$. If we take eigenvectors U_j, $j = r + 1, \ldots, L$, as a basis of $\mathcal{L}_r^\perp$, $\mathbf{U}_{r+1,L} = [U_{r+1} : \ldots : U_L]$, then $\mathbf{U}_{r+1,L}\mathbf{U}_{r+1,L}^*$, where $*$ denotes conjugate transposition, provides the matrix of projection on $\mathcal{L}_r^\perp$ and therefore $f(\omega) = W^*(\mathbf{U}_{r+1,L}\mathbf{U}_{r+1,L}^*)W/\|W\|^2 = \sum_{j=r+1}^{L} f_j(\omega)$, where $f_j(\omega) = \cos^2(\widehat{W(\omega), U_j})$. Thus, the MUSIC method can be considered from the viewpoint of the subspace properties and does not require the computation of the roots of the characteristic polynomials. Similar to the Min-Norm method, the MUSIC method is essentially the method of searching for the maximums of the pseudospectrum $1/f(\omega)$.

3.8.4 Hankel SLRA

3.8.4.1 Cadzow Iterations

The aim of Cadzow iterations is to extract the finite-rank signal $\mathbb{S}$ of rank r from an observed noisy signal $\mathbb{X} = \mathbb{S} + \mathbb{P}$. Cadzow iterations [8] were suggested as a method of signal processing, without any relation to SSA method. However, these two methods are very much related.

Cadzow iterations present an example of the procedure called alternating projections. A short form of M iterations is

$$\widetilde{\mathbb{S}} = \mathcal{T}^{-1} \circ \left(\boldsymbol{\Pi}_{\mathcal{H}} \circ \boldsymbol{\Pi}_r\right)^M \circ \mathcal{T}(\mathbb{X}). \tag{3.24}$$

Here the embedding operator $\mathcal{T}$ provides a one-to-one correspondence between time series and trajectory matrices for the fixed window length L, $\boldsymbol{\Pi}_r$ is the projection of a matrix to the space $\mathcal{M}_r$ of $L \times K$ matrices of rank not larger than r, the hankelisation operator $\boldsymbol{\Pi}_{\mathcal{H}}$ is also the projection into the space $\mathcal{H}$ of Hankel matrices in the Frobenius norm. See how to call the corresponding R functions from the RSSA package in [24, Sect. 3.4.3].

The Basic SSA with fixed L and fixed grouping $I = \{1, 2, \ldots, r\}$ is simply the first iteration of Cadzow iterations, see (2.22). This means that Cadzow iterations can be defined as a repeated application of Basic SSA with fixed L and fixed grouping $I = \{1, 2, \ldots, r\}$ to the series $\widetilde{\mathbb{X}}_I$, see (2.7), obtained by Basic SSA in the previous step; the initial Cadzow iteration is Basic SSA applied to the original series $\mathbb{X}$.

The result of Cadzow iterations is a signal of finite rank $\leq r$. However, this does not guarantee that the limiting result is closer to the true signal than SSA result (that is, just one iteration). Among other factors, this depends on how well the true signal can be approximated by the series of rank r and the recommended choice of L: indeed, a usual recommendation in signal processing literature is to choose L which is just slightly larger than r; this, however, is unwise from the viewpoint of SSA.

Cadzow iterations are easily extended to the multidimensional case. Cadzow iterations can also be generalized to weighted iterations in a natural way, when the projectors in (3.24) are constructed with respect to a given weighted norm.

There are weights, when the implementation of the weighted projections has computational cost comparable with that in the unweighted case [50]: the inner product in $\mathsf{R}^{L \times K}$ is defined as $\langle \mathbf{X}, \mathbf{Y} \rangle = \langle \mathbf{X}, \mathbf{Y} \rangle_{\mathbf{L}, \mathbf{R}} = \mathrm{tr}(\mathbf{LXRY}^\mathsf{T})$, where $\mathbf{L} \in \mathsf{R}^{L \times L}$, $\mathbf{R} \in \mathsf{R}^{K \times K}$; $\|\mathbf{X}\|_{\mathbf{L}, \mathbf{R}}$ is the corresponding matrix norm in $\mathsf{R}^{L \times K}$. Below in this section we will explain the problem of choice of the weights for more accurate estimates.

Formally, Cadzow iterations constitute a method of solving the general HSLRA (Hankel-structured matrix low-rank approximation) problem considered next. The series $\widetilde{\mathbb{S}}_N$ obtained by (3.24) can be regarded as an estimator of the signal. Note, however, that Cadzow iterations may not converge even to a local optimum of the respective optimization problem (3.26).

3.8.4.2 General HSLRA Problem

We will formulate the HSLRA problem as the problem of extraction of a signal $\mathbb{S} = (s_1, s_2, \ldots, s_N)^\mathsf{T}$ of rank r from an observed noisy signal $\mathbb{X} = (x_1, x_2, \ldots, x_N)^\mathsf{T} = \mathbb{S} + \mathcal{E}$ of length N, where $\mathcal{E} = (\epsilon_1, \epsilon_2, \ldots, \epsilon_N)^\mathsf{T}$ is a vector of (unobserved) random noise with zero mean and covariance matrix $\mathbf{\Sigma} = \mathsf{E}\mathcal{E}\mathcal{E}^\mathsf{T}$; see also [9, 34, 50].

The HSLRA problem can be stated in two forms: (a) vector form and (b) matrix form. The vector (time series) form is: for given $\mathbb{X} \in \mathsf{R}^N$ and positive integer $r < \lfloor N/2 \rfloor$,

$$\|\mathbb{X} - \mathbb{Y}\|_{\mathbf{W}}^2 \to \min_{\mathbb{Y}:\mathrm{rank}\, \mathbb{Y} \leq r} \tag{3.25}$$

where the matrix $\mathbf{W} \in \mathsf{R}^{N \times N}$ is positive definite and $\|\mathbb{Z}\|_{\mathbf{W}}^2 = \mathbb{Z}\mathbf{W}\mathbb{Z}^\mathsf{T}$ for a row-vector $\mathbb{Z}$. The solution of (3.25) can be considered as a weighted least-squares estimate (WLSE) of the signal $\mathbb{S}$. If noise $\mathcal{E}$ is Gaussian with covariance matrix $\mathbf{\Sigma}$, then the WLSE with $\mathbf{W} = \mathbf{\Sigma}^{-1}$ is the maximum likelihood estimate (MLE). If the properties of the noise process are known then the vector form (3.25) is the most natural way of defining the HSLRA problem. However, solving the HSLRA problem in the vector form is extremely difficult, see e.g. [12]. Although the vector form allows fast implementations, these implementations are very complex and need a starting point close to the solution [35, 51].

The matrix form of the HSLRA problem allows one to use simple subspace-based alternating projection methods like Cadzow iterations considered above. It is the following optimization problem:

$$\|\mathbf{X} - \mathbf{Y}\|_{\mathbf{L,R}}^2 \to \min_{\mathbf{Y} \in \mathcal{M}_r \cap \mathcal{H}}. \tag{3.26}$$

For reformulating the original HSLRA problem (3.25) in the matrix form (3.26), we have to choose $\mathbf{X} = \mathcal{T}_L(\mathbb{X})$ and $\mathbf{Y} = \mathcal{T}_L(\mathbb{Y})$; the remaining issue is matching the vector norm in (3.25) with the matrix norm in (3.26).

The general case of the correspondence between the vector-norm and matrix-norm formulations (3.25) and (3.26) of the HSLRA problem is established in the following theorem [13, 20]; $*$ means the matrix convolution.

Theorem 3.2 *For any $\mathbb{Z} \in \mathsf{R}^N$, $\|\mathcal{T}_L(\mathbb{Z})\|_{\mathbf{L,R}} = \|\mathbb{Z}\|_{\mathbf{W}}$ if and only if*

$$\mathbf{W} = \mathbf{L} * \mathbf{R}. \tag{3.27}$$

In a typical application, when the structure of the noise in the model 'signal plus noise' is assumed, the HSLRA problem is formulated in a vector form with a given matrix $\mathbf{W}$. As mentioned above, algorithms of solving the HSLRA problem are much easier if we have the matrix rather than vector form of the HSLRA problem. Therefore, in view of Theorem 3.2, for a given $\mathbf{W}$ we would want to find positive definite matrices $\mathbf{L}$ and $\mathbf{R}$ such that (3.27) holds; that is, we would want to perform a blind deconvolution of the matrix $\mathbf{W}$.

It follows from the results of [49] that in the case when the noise $\mathcal{E}$ is white, and therefore $\mathbf{W} = \mathbf{I}_N$, the matrix $\mathbf{W}$ cannot be blindly deconvoluted under the condition that $\mathbf{L}$ and $\mathbf{R}$ are positive definite matrices. The paper [20] extends the results of [49] to the case of banded matrices corresponding to the case where the noise $\mathcal{E}$ forms an autoregressive process.

Non-existence of symmetric positive definite weight matrices in the matrix form of Hankel SLRA means that there is no matrix form of the Hankel SLRA problem, which corresponds to the weighted least-squares problem in vector form with optimal weights. However, we can reasonably simply find an approximate solution to the matrix equation (3.27), see [50].

3.9   SSA and Filters

As demonstrated in Sect. 2.3, one of SSA's capabilities is its ability to be a frequency filter. The relation between SSA and filtering was considered in a number of papers, see for example [6, 28]. These results are mostly related to the case where (a) the window length L is small (much less than $N/2$), and (b) Toeplitz SSA is considered and the filter properties are based on the properties of the eigenvectors of Toeplitz matrices (therefore, the time series is assumed to be stationary, see Sect. 2.5.2.2).

In this section we describe the relation between Basic SSA and filtering in a general form and also consider specific filters generated by Basic SSA.

3.9.1 Linear Filters and Their Characteristics

Let $\mathbf{x} = (\ldots, x_{-1}, \overset{\circ}{x_0}, x_1, x_2, \ldots)$ be an infinite sequence and the symbol 'o' over an element denotes its middle location. Finite series $\mathbb{X}_N = (x_1, \ldots, x_N)$ can be formally presented as a infinite sequence $(\ldots, 0, \ldots, \overset{\circ}{0}, x_1, x_2, \ldots, x_N, 0, \ldots)$. Each linear filter Φ can be expressed as $\big(\Phi(\mathbf{x})\big)_j = \sum_{i=-\infty}^{+\infty} h_i x_{j-i}$. The sequence $\mathbf{h}_\Phi = (\ldots, h_{-1}, \overset{\circ}{h_0}, h_1, \ldots)$ is called *the impulse response*. A filter Φ is called FIR-filter (i.e. with Finite Impulse Response) if $\big(\Phi(\mathbf{x})\big)_j = \sum_{i=-r_1}^{r_2} h_i x_{j-i}$. The filter Φ is called *causal* if $\big(\Phi(\mathbf{x})\big)_j = \sum_{i=0}^{r-1} h_i x_{j-i}$.

The following characteristics are standard for filters: $H_\Phi(z) = \sum_i h_i z^{-i}$ is a *transfer function*, $A_\Phi(\omega) = |H_\Phi(e^{i2\pi\omega})|$ is a *frequency (amplitude) response* and $\varphi_\Phi(\omega) = \operatorname{Arg} H_\Phi(e^{i2\pi\omega})$ is a *phase response*. The meaning of the amplitude and phase responses follows from: for the sequence $\mathbf{x}$ with $(\mathbf{x})_j = \cos(2\pi\omega j)$ we have $\big(\Phi(\mathbf{x})\big)_j = A_\Phi(\omega)\cos(2\pi\omega j + \varphi_\Phi(\omega))$.

An important filter characteristic reflecting its noise reduction capability is the filter *power* $\mathscr{E}\Phi = \|\mathbf{h}\|^2 = \sum_i h_i^2$. The following proposition is analogous to Proposition 3.4.

Proposition 3.8 *Let* $\mathbf{x} = \mathbf{s} + \varepsilon$, *where* $(\varepsilon)_j$ *are i.i.d,* $\mathsf{E}(\varepsilon)_j = 0$, $\mathsf{D}(\varepsilon)_j = \sigma^2$. *Let* $\Phi\colon \Phi(\mathbf{s}) = \mathbf{s}$ *and denote* $\widetilde{\mathbf{x}} = \Phi(\mathbf{x})$. *Then* $\mathsf{E}(\widetilde{\mathbf{x}})_j = (\mathbf{s})_j$ *and* $\mathsf{D}(\widetilde{\mathbf{x}})_j = \sigma^2 \cdot \mathscr{E}\Phi$.

Also, there is a relation between the filter power and the frequency response. Define $\Delta_a\Phi = \operatorname{meas}\{\omega \in (-0.5, 0.5]: A_\Phi(\omega) \geq a\}$. Parseval's identity has the following form for filters:

$$\mathscr{E}\Phi = \sum_j h_j^2 = \int_{-0.5}^{0.5} A_\Phi(\omega)^2 \, d\omega.$$

Therefore,

$$\Delta_a\Phi \leq \mathscr{E}\Phi/a^2. \tag{3.28}$$

The inequality (3.28) shows how the support of the frequency response (with threshold a) is related to the filter power.

3.9.2 SSA Reconstruction as a Linear Filter

Let us return to Basic SSA. Let L be the window length and $(\sqrt{\lambda}, U, V)$ be one of the eigentriples generated by the SVD of the trajectory matrix of $\mathbb{X}_N$ (see Sect. 2.1.1 for notation and definitions). Since the reconstruction operation in Basic SSA is the linear operation, it can be written in matrix form.

Let $K = N - L + 1$, $L^* = \min(L, K)$. Define the diagonal $N \times N$ matrix

$$\mathbf{D} = \mathrm{diag}(1, 2, 3, \dots, L^* - 1, L^*, L^*, \dots, L^*, L^* - 1, \dots, 2, 1)$$

and the $K \times N$ matrix

$$\mathbf{W} = \begin{pmatrix} u_1 & u_2 & u_3 & \cdots & u_L & 0 & \cdots & 0 & 0 & 0 \\ 0 & u_1 & u_2 & u_3 & \cdots & u_L & 0 & \cdots & 0 & 0 \\ \vdots & 0 & \ddots & \ddots & \ddots & \cdots & \ddots & 0 & \cdots & 0 \\ 0 & \cdots & 0 & u_1 & u_2 & u_3 & \cdots & u_L & 0 & \vdots \\ 0 & 0 & \cdots & 0 & u_1 & u_2 & u_3 & \cdots & u_L & 0 \\ 0 & 0 & 0 & \cdots & 0 & u_1 & u_2 & u_3 & \cdots & u_L \end{pmatrix}.$$

Proposition 3.9 *The time series component $\widetilde{\mathbb{X}}_N$ reconstructed by the eigentriple $(\sqrt{\lambda}, U, V)$ has the form*

$$\widetilde{\mathbb{X}}_N^{\mathrm{T}} = \mathbf{D}^{-1} \mathbf{W}^{\mathrm{T}} \mathbf{W} \mathbb{X}_N^{\mathrm{T}}.$$

Proof First, note that $\mathbf{W} \mathbb{X}_N^{\mathrm{T}} = \mathbf{X}^{\mathrm{T}} U = \sqrt{\lambda} V \in \mathsf{R}^K$. This yields that the vector $\mathbf{W}^{\mathrm{T}} \mathbf{W} \mathbb{X}_N^{\mathrm{T}}$ (of size N) consists of sums along N antidiagonals of the matrix $\sqrt{\lambda} U V^{\mathrm{T}}$, which is an elementary summand of the SVD. Multiplication by $\mathbf{D}^{-1}$ provides the normalization of the sums by the number of summands and therefore by the definition we obtain the elementary reconstructed component. □

Remark 3.11 Let us add the index i to $\widetilde{\mathbb{X}}$ to indicate that it corresponds to the i-th eigenvector $U = U_i$. Then, evidently, the reconstructed series $\widetilde{\mathbb{X}}^{(I)}$ by the set of eigentriples $\{(\sqrt{\lambda_i}, U_i, V_i), \ i \in I\}$ is equal to the sum of the reconstructed elementary series $\widetilde{\mathbb{X}}^{(i)}$. Therefore, the matrix form for $\widetilde{\mathbb{X}}^{(I)}$ immediately follows from (3.9).

Proposition 3.9 and Remark 3.11 allow us to describe the reconstruction produced by Basic SSA as an application of a set of linear filters.

Let $L \leq K$. Define the linear filters $\Theta_L, \Theta_{L-1}, \dots, \Theta_1$ and Ψ by their impulse characteristics $\mathbf{h}_{\Theta_L}, \dots, \mathbf{h}_{\Theta_1}$ and $\mathbf{h}_\Psi$:

$$\mathbf{h}_{\Theta_L} = (\dots, 0, \overset{\circ}{u}_L, 0, \dots),$$
$$\mathbf{h}_{\Theta_{L-1}} = (\dots, 0, u_{L-1}, \overset{\circ}{u}_L, 0, \dots),$$
$$\dots$$
$$\mathbf{h}_{\Theta_1} = (\dots, 0, u_1, \dots, u_{L-2}, u_{L-1}, \overset{\circ}{u}_L, 0, \dots);$$
$$\mathbf{h}_\Psi = \operatorname{rev} \mathbf{h}_{\Theta_1} = (\dots, 0, \overset{\circ}{u}_L, u_{L-1}, \dots, u_1, 0, \dots).$$

Now we can introduce the reconstructing filters $\Phi_k, k = 1, \dots, L$, generated by the vector U:

$$\Phi_k = \Theta_k \circ \Psi / (L - k + 1), \tag{3.29}$$

where 'o' stands for the filter composition, which is equivalent to the convolution '$*$' of the filter impulse characteristics.

Proposition 3.10 *For* $\mathbb{X}_N = (x_1, x_2, \dots, x_N)$, *the terms of the elementary reconstructed series* $\widetilde{\mathbb{X}}_N$ *corresponding to the eigentriple* $(\sqrt{\lambda}, U, V)$ *have the following representation:*

- $\widetilde{x}_s = (\Phi_{s-K+1}(\mathbb{X}_N))_s$ *for* $K + 1 \le s \le N$;
- $\widetilde{x}_s = (\Phi_1(\mathbb{X}_N))_s$ *for* $L \le s \le K$.

This result is a direct consequence of Proposition 3.9. The set of filters providing the reconstruction of $\widetilde{x}_s$ for $1 \le s < L$ can be built in a similar way.

Let us examine two special filters: Φ_1, which is used for the reconstruction of the middle points of the time series with numbers $L, \dots, K$, and Φ_L, which is used for the reconstruction of the last point only. The former is called *the MPF* (Middle Point Filter) and the latter is referred as *the LPF* (Last Point Filter). In next two sections we consider them separately.

3.9.3 Middle Point Filter

As above, we assume $L \le K$. According to Proposition 3.10, the MPF Φ_1 acts only at the L-th to $(N-L+1)$-th points. This leads to a limited use of the MPF in the case $L \sim N/2$. The explicit formula for the MPF filter $\Phi_1^{(i)}$ corresponding to the eigenvector $U_i = (u_1, \dots, u_L)^\mathrm{T}$ has the following form:

$$\widetilde{x}_s = \sum_{j=-(L-1)}^{L-1} \left(\sum_{k=1}^{L-|j|} u_k u_{k+|j|} / L \right) x_{s-j}, \quad L \le s \le K. \tag{3.30}$$

It is clearly seen that the order of the MPF is equal to $2L - 1$. Alternative representation of (3.30) is

$$\tilde{x}_s = \sum_{j=1}^{L} \sum_{l=1}^{L} u_j u_l x_{s+j-l}/L, \ L \leq s \leq K. \tag{3.31}$$

Let us enumerate several properties of the MPF $\Phi_1^{(I)}$.

1. The filter $\Phi_1^{(I)}$ is symmetric. Hence the MPF is a zero-phase filter. In particular, the MPF does not change phases of sinusoids.
2. In a particular case of $I = \{i\}$, applying Jensen's inequality, we obtain that the sum of coefficients of the $\Phi_1^{(i)}$ given in (3.31) is not larger than 1:

$$\sum_{j=1}^{L} \sum_{l=1}^{L} u_j u_l/L = \left(\sum_{j=1}^{L} u_j\right)^2 \bigg/ L \leq \sum_{j=1}^{L} u_j^2 = 1.$$

3. If the matrix $\mathbf{XX}^T$ is positive, then the leading eigenvector U_1 is positive too (Perron's theorem) and, therefore, the coefficients of the filter $\Phi_1^{(1)}$ are positive. This is true, for example, in the case of positive time series. If the time series is close to a constant (at the timescale of L), then the coordinates of U_1 will be close one to another and the MPF filter $\Phi_1^{(1)}$ will be close to the so-called triangle (Bartlett) filter. This implies, for instance, that the extraction of trend by the first run of Sequential SSA with small L (see Sect. 2.5.4) is similar to the application of a weighted moving average procedure with positive nearly triangular weights.
4. Power of the MPF satisfies the following inequalities.

Proposition 3.11 Let the filter $\Phi_1^{(i)}$ be the MPF generated by an eigenvector $U_i = (u_1, \ldots, u_L)^T$. Then its power satisfies the inequality $\mathscr{E}\Phi_1^{(i)} \leq 1/L$.

Proof The proof of the proposition results from the following inequality:

$$\|\mathbf{h}_\Psi * \mathrm{rev}\,\mathbf{h}_\Psi\| \leq \sum_{j=1}^{L} |u_j| \cdot \|\mathbf{h}_\Psi\| = \sum_{j=1}^{L} |u_j| \cdot \|U\| = \sum_{j=1}^{L} |u_j| \leq \sqrt{L}\|U\| = \sqrt{L}.$$

$\square$

Proposition 3.12 Let $\Phi_1^{(I)}$ be the MPF generated by eigenvectors $\{U_i, i \in I\}$ where $|I| = r$. Then its power satisfies the inequality $\mathscr{E}\Phi_1^{(I)} \leq r^2/L$.

Proof By the linearity of the grouping operation, $\Phi_1^{(I)} = \sum_{i \in I} \Phi_1^{(i)}$, and therefore, by Proposition 3.11 we have:

$$\mathscr{E}\Phi_1^{(I)} = \left\|\sum_{i \in I} \mathbf{h}_{\Phi_1^{(i)}}\right\|^2 \leq \left(\sum_{i \in I} \left\|\mathbf{h}_{\Phi_1^{(i)}}\right\|\right)^2 = \left(\sum_{i \in I} \sqrt{\mathscr{E}\Phi_1^{(i)}}\right)^2 \leq r^2/L.$$

$\square$

5. A direct consequence of Proposition 3.12 and inequality (3.28) is the inequality $\Delta_a \Phi_1^{(I)} \le r^2/(a^2 L)$. This means that for any threshold a, the support of filter frequency response tends to 0 as $L \to \infty$. This effect is clearly seen in Fig. 2.22 (Sect. 2.4.3) showing the smoothing effect of Basic SSA.

6. Let us define for $\omega \in (-0.5, 0.5]$:

$$g_U(\omega) = \frac{1}{L} \left| \sum_{j=1}^{L} u_j e^{-i2\pi\omega j} \right|^2. \tag{3.32}$$

The function g_U is closely related to the periodogram Π_u^L introduced in (2.10) of Sect. 2.3.1.1: $g_U(k/L) = L \, \Pi_u^L(k/L)/2$ for $0 < k < N/2$ and $g_U(k/L) = L \, \Pi_u^L(k/L)$ otherwise. It appears that the frequency response of the MPF is almost the same as the periodogram of the vector U.

Proposition 3.13 Let A_{Φ_1} be the frequency response of the MPF filter Φ_1. Then $g_U(\omega) = A_{\Phi_1}(\omega)$.

Proof Recall that $\Phi_1 = \Theta_1 \circ \Psi/L$, where $\mathbf{h}_\Psi = (\ldots, 0, \mathring{u}_L, u_{L-1}, \ldots, u_1, 0, \ldots)$ and $\mathbf{h}_{\Theta_1} = \operatorname{rev} \mathbf{h}_\Psi$. Also, from the theory of linear filters [37] we have $A_{\Phi_1 \circ \Psi}(\omega) \equiv A_{\Phi_1}(\omega) A_\Psi(\omega)$. Then

$$A_{\Phi_1}(\omega) = \frac{1}{L} \left| \sum_{j=0}^{L-1} u_{L-j} e^{-i2\pi\omega j} \right| \cdot \left| \sum_{j=0}^{1-L} u_{L+j} e^{-i2\pi\omega j} \right| = \frac{1}{L} \left| \sum_{j=1}^{L} u_j e^{-i2\pi\omega j} \right|^2.$$

$\square$

7. It follows from Proposition 3.13 that for SSA identification and interpretation of the SVD components, the periodogram analysis of eigenvectors can be very helpful. Also, an automatic identification of components introduced in Sect. 2.4.5 is based on properties of periodograms of eigenvectors and therefore can also be expressed in terms of the frequency response of the MPF.

3.9.4 Last Point Filter and Forecasting

The last-point filter (LPF) is not really a filter as it is used only for the reconstruction of the last point: $\widetilde{x}_N = \sum_{i=0}^{L-1} u_L u_{i+1} x_{N-i}$. The reconstruction by the eigentriples with numbers from the set I has the following form:

$$\widetilde{x}_N^{(I)} = \sum_{k=0}^{L-1} \left(\sum_{i \in I} u_L^{(i)} u_{k+1}^{(i)} \right) x_{N-k}. \tag{3.33}$$

However, it is the only reconstruction filter that is causal. This has two consequences. First, the LPF of a finite-rank series is closely related to the LRR governing and forecasting this time series. Second, the so-called Causal SSA (or last-point SSA) can be constructed by means of the use of the last reconstructed points of the accumulated data. Since in Causal SSA the LPF is applied many times, studying properties of LPF is important.

Let the signal $\mathbb{S}_N$ has rank r and is governed by an LRR. Unbiased causal filters of order L and linear recurrence relations of order $L - 1$ are closely related. In particular, if the causal filter is given by $s_j = \sum_{k=0}^{L-1} a_{L-k} s_{j-k}$ and $a_L \neq 1$, then this filter generates the following LRR of order $L - 1$: $s_j = \sum_{k=1}^{L-1} c_{L-k} s_{j-k}$, where $c_k = a_k/(1 - a_L)$.

Similar to the minimum-norm LRR, the minimum-power filters can be considered. It appears that the LPF has minimal power among all unbiased filters. This follows from the relation between LRRs and causal filters. Denote by $\mathbf{P}_r$ the orthogonal projector onto the signal subspace. The LPF has the form $s_N = (\mathbf{P}_r S_K)_L = A^{\mathrm{T}} S_K$, where $A = \mathbf{P}_r \mathbf{e}_L$, while the min-norm LRR is produced by the last-point filter, i.e. $R = \underline{A}/(1 - a_L)$.

In the general case, the filter (3.33) can be rewritten as $\widetilde{x}_N = (\mathbf{P}_r X_K)_L$, where X_K is the last L-lagged vector, $\mathbf{P}_r$ is the projector on the SSA estimate of the signal subspace $\mathrm{span}(U_i, i \in I)$. Formally applying this filter to the whole time series, we obtain the series of length K consisting of the last points of the reconstructed lagged vectors $\widetilde{X}_k$.

Note that if we use other estimates of the signal subspace, then we obtain other versions of the last-point filter.

3.9.5 Causal SSA (Last-Point SSA)

Let $\mathbb{X}_\infty = (x_1, x_2, \ldots)$ be an infinite series, $\mathbb{X}_M = (x_1, \ldots, x_M)$ be its subseries of length M, L be fixed, $\mathcal{L}(M)$ be a subspace of R^L, $\mathrm{P}(M)$ be a projector to $\mathcal{L}(M)$, $A(M) = \mathrm{P}_r(M) \mathbf{e}_L$, $K = K(M) = M - L + 1$.

Introduce the series $\check{\mathbb{X}}_\infty$ as follow. Define $(\check{\mathbb{X}}_\infty)_M = (\mathrm{P}(M) X_{K(M)})_L = A(M)^{\mathrm{T}} X_{K(M)}$, where $X_{K(M)}$ is the last L-lagged vector of $\mathbb{X}_M$. Thus, $(\check{\mathbb{X}}_\infty)_M$ is a linear combination of the last L terms of $\mathbb{X}_M$ with coefficients depending on M; that is, $\check{\mathbb{X}}_\infty$ can be considered as a result of application of a sequence of different causal filters to $\mathbb{X}_\infty$.

If $\mathcal{L}_r(M) = \mathcal{L}(M) = \mathrm{span}(U_1(M), \ldots, U_r(M))$, where $U_1(M), \ldots, U_r(M)$ are the signal eigenvectors produced by SSA with window length L applied to $\mathbb{X}_M$, then this sequence of causal filters is called Causal SSA. In this case, $(\widetilde{\mathbb{X}})_M$ is equal to the last point of SSA reconstruction $\widetilde{\mathbb{X}}_M$, which in turn is equal to the last coordinate of the projection of the last lagged vector of $\mathbb{X}_M$ to $\mathcal{L}_r(M)$.

Given that $\mathcal{L}_r(M)$ is used as an estimate of the signal subspace, M should be large enough to provide a good estimate. Therefore, we need to introduce a starting point M_0 (with $M_0 > L$) in Causal SSA and consider $M \geq M_0$ only.

Since the result of application of Causal SSA is a sequence $\check{\mathbb{X}}_\infty$ built from SSA reconstructions of the last points of the subseries, Causal SSA can be called Last-point SSA.

Remark 3.12 Let us fix N, L and consider $\mathbb{X}_N$, the corresponding $U_1, \ldots, U_r$ and $\mathcal{L}_r(M) = \mathrm{span}(U_1, \ldots, U_r)$ for any M. Then the series $\check{\mathbb{X}}_N$ is the result of application of the last-point filter to $\mathbb{X}_N$. Assuming that the estimates of the signal subspace on the base of $\mathbb{X}_M$ are stable for large enough M, we can conclude that the result of Causal SSA will be close to the result of the last-point filter (LPF) applied to the whole series.

Note also that in the considered particular case, $\check{\mathbb{X}}_N$ (more precisely, its last K points; the first $L - 1$ points are not defined or can be set to zero) coincides with the last row of the reconstructed matrix $\widehat{\mathbf{X}}$ of the series $\mathbb{X}_N$. That is, $\check{\mathbb{X}}_N$ is similar to the result of SSA reconstruction before the diagonal averaging is made.

Causality yields the following relation: under the transition from $\mathbb{X}_M$ to $\mathbb{X}_{M+1}$, the first M points of the $\check{\mathbb{X}}_{M+1}$ coincide with $\check{\mathbb{X}}_M$. This is generally not true if we consider the reconstructions $\widetilde{\mathbb{X}}_M$ for $\mathbb{X}_M$ obtained by the conventional SSA. The effect $(\widetilde{\mathbb{X}}_M)_j \neq (\widetilde{\mathbb{X}}_{M+1})_j$, $j \leq M$, is called 'redrawing'. For real-life time series we usually have redrawing for all j and the amount of redrawing depends on j. Redrawing of only a few last points is usually of interest. Moreover, redrawing of local extremes of the series is more practically important than redrawing of regular series points. Small values of redrawing indicate stability of SSA decompositions and hence stability of time series structure. The amount of redrawing can be assessed by visual examination or measured using the variance of the redrawings at each time moment of interest. These variances can be averaged if needed.

Generally, the reconstruction has no delay (at least, the middle-point filter has zero phase shift). On the other hand, delays in Causal SSA are very likely. In a sense, a redrawing in SSA is converted to a delay in Causal SSA. If $\mathcal{L}_r$ corresponds to an exactly separated component of the time series $\mathbb{X}$, then SSA has no redrawing and Causal SSA has no delay. In conditions of good approximate separability the redrawing is almost absent and the delay is small.

Example Let us demonstrate the described effects on the 'S&P500' example introduced in Sect. 2.5.1. Figure 3.7 shows the result of Causal SSA with window length $L = 30$, $\mathcal{L}_r(M) = \mathrm{span}(U_1(M), U_2(M))$ and $M_0 = 200$. The delay is clearly seen. If we consider non-causal (Basic) SSA reconstructions of cumulative subseries, then the redrawing takes place (Fig. 3.8). This redrawing increases in the points of local maximums and minimums.

Finally, let us note that if the direction of change of the time series is of primary concern (rather than the values themselves), then, instead of taking differences of Causal SSA series, it may be worthwhile considering the time series which consists of differences of the last two points of reconstructions. The result should be expected to be more stable since the reconstruction of the last-but-one point has better accuracy than that of the last point.

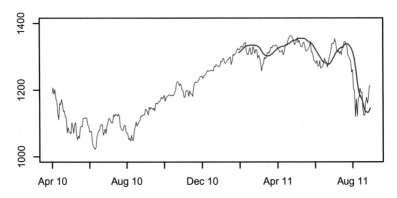

Fig. 3.7 S&P500: Causal SSA

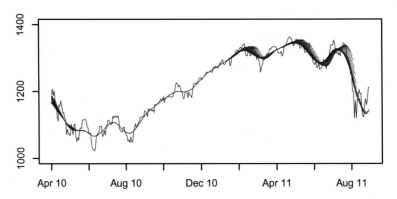

Fig. 3.8 S&P500: Non-causal SSA with redrawing

3.10   Multidimensional/Multivariate SSA

In Sect. 2.6, we briefly discussed the approach to construction of multidimensional extensions of SSA for decomposing the initial object into identifiable components. This general approach to SSA, developed in [23, 24], makes it easy to extend the algorithms of SSA-related methods developed for the analysis of one-dimensional (1D) time series to the cases of collections of time series and digital images.

This concerns the approaches to decomposition, refined decompositions including rotations, filtering, parameters estimation, forecasting, gap filling and also to the choice of parameters. In this section, we dwell on some differences between the 1D and multidimensional cases.

The main notions in SSA are signal subspace, separability, rank, LRR. Their definitions are the same for different dimensionalities.

The trajectory space is constructed as the span of columns or rows of the trajectory matrix. Signal subspace is defined as the trajectory space of the signal of interest. The (weak) separability is determined exactly in the same way as in the 1D case via orthogonality of the trajectory spaces of time series components which should be separated. The strong separability is determined as non-coincidence of the singular values of trajectory matrices of these components.

Rank of a multidimensional object is defined as rank of its trajectory matrix. If for all sufficiently large window sizes and object sizes: (a) the trajectory matrix is rank-deficient then the object is called rank deficient, and (b) the ranks of the trajectory matrices equal r then we say that the object has finite rank r.

SSA forecasting and gap filling are also similar to the 1D case: first, the signal subspace is estimated and then the imputation of unknown data is performed to keep the signal subspace.

The objects of finite rank produce a class of objects, which satisfy LRRs. Again, the objects satisfying LRRs have the parametric form of a finite sum of products of multivariate polynomials, exponentials and sinusoids (see e.g. [18, Proposition 7] and [23, Appendix B]). LRRs correspond to characteristic polynomials of several variables, whose roots are needed for estimation of parameters.

Although both 2D-SSA and MSSA comply with these principles (moreover, MSSA can be considered as a particular case of 2D-SSA), there are significant differences. For 2D objects, we consider both dimensions from the same viewpoint; in particular, the words 'sufficiently large' are related to both sizes of the object and the window in both dimensions. For multivariate time series, the number of time series in the collection is fixed and the window size with respect to this dimension is equal to 1.

3.10.1 MSSA

Despite formally MSSA is a particular case of 2D-SSA, from the viewpoint of analysis, MSSA is closer to 1D-SSA than to 2D-SSA. In particular, finite-rank collections of time series are described in terms of the collections of time series, where each series satisfies an LRR. Thus, we will not discuss the common things that have been thoroughly discussed for time series.

Ranks and common structure of time series. In MSSA analysis, it is important if all time series in the chosen collection have similar behaviour; if these time series are different, then it is better to analyse them separately. In the context of SSA, similar behaviour means a common structure of signal subspaces. From the viewpoint of LRRs, the characteristic polynomials of different series from the collection have the same (or with a large intersection) set of roots. For instance, sinusoids with the same periods have the same structure; the same is true for linear trends.

An indicator of the same structure is the relation between the MSSA rank r of the collection of signals and the SSA ranks r_i of signals, $i = 1, \ldots, s$. The same structure of all signals corresponds to the case $r_1 = \cdots = r_s = r$, while entirely different structures yield the equality $r = \sum_{i=1}^{s} r_i$.

Within the framework of SSA, we cannot talk about causality, which implies time shifts, since the MSSA method is invariant with respect to shifts in time. However, we can talk about the supportiveness of one series with respect to another series. The supportiveness of the second time series with respect to the first time series means that the second time series improves the accuracy of signal estimation or forecasting in comparison with the use of the first series only. The supportiveness depends on the similarity of the signals subspaces and on noise levels. Note that if one time series is supportive for another one, this does not imply that the second time series is supportive for the first one. An easy example illustrating this is: both time series consist of the same signal $s_n = \cos(2\pi \omega n)$ corrupted by different noises, very small and very large, respectively.

Forecasting. We have mentioned in Sect. 2.6.1 that for MSSA, the rows and columns of the trajectory matrix have different forms. Therefore, the methods of MSSA forecasting require special attention.

As in 1D-SSA, methods of MSSA forecasting can be subdivided into recurrent and vector forecasting. In contrast with 1D-SSA, there exist two kinds of MSSA forecasting: row forecasting and column forecasting; this depends on which of the two spaces the forecasting is made (row or column space respectively). In total, there are four main variants of MSSA forecasting: recurrent column forecasting, recurrent row forecasting, vector column forecasting and vector row forecasting.

Note that there are several different names for the same SSA forecasting methods, see [17, 23, 29]. In Sect. 2.6.1, we have explained the choice of orientation of the MSSA trajectory matrix and the connection between the horizontally-stacked and vertically-stacked trajectory matrices of separate time series. We follow [23, 24] and use the name 'column' and 'row' with respect to the horizontally-stacked trajectory matrices as defined in Sect. 2.6.1. In the column forecasting methods, each time series in the collection is forecasted separately in a given common subspace (that is, using a common LRR). In the row forecasting methods, each series is forecasted with the help of its own LRR applied to the whole set of series from the collection. The capabilities of the RSSA package for MSSA forecasting are discussed in [24, Sect. 4.3.3].

Missing data imputation, parameter estimation and other subspace-based methods are performed in the same manner as for 1D-SSA.

Numerical comparison of forecasts for a simulated data. In [17, 23, 29], a comparison was performed for series without trends. Here we add a linear trend to both time series. Let us assume that we observe $(\mathbb{X}^{(1)}, \mathbb{X}^{(2)}) = (\mathbb{S}^{(1)}, \mathbb{S}^{(2)}) + (\mathbb{N}^{(1)}, \mathbb{N}^{(2)})$, where $(\mathbb{S}^{(1)}, \mathbb{S}^{(2)})$ is a two-dimensional signal, $\mathbb{N}^{(1)}$ and $\mathbb{N}^{(2)}$ are realizations of independent white Gaussian noises. Then we can use the standard simulation techniques to obtain estimates of the mean square errors (MSE) for the reconstruction and forecasting of $(\mathbb{S}^{(1)}, \mathbb{S}^{(2)})$ by the indicated SSA methods. The resultant MSE is calculated as the mean of $\text{MSE}^{(1)}$ and $\text{MSE}^{(2)}$ for $\mathbb{S}^{(1)}$ and $\mathbb{S}^{(2)}$ correspondingly. We take the following parameters for the simulation of the time series: $N = 71$, the variance of each noise components is $\sigma^2 = 25$, the number of replications is 1000:

$$s_k^{(1)} = 30\cos(2\pi k/12) + k, \quad s_k^{(2)} = 30\cos(2\pi k/12 + \pi/4) + 100 - k, \quad (3.34)$$

$$k = 1, \ldots, N.$$

Table 3.4 MSE of signal forecasts

Example (3.34)	$L = 12$	$L = 24$	$L = 36$	$L = 48$	$L = 60$
Recurrent					
MSSA-column	1879.08	32.15	**20.55**	21.70	22.44
MSSA-row	177.20	34.02	23.54	**19.88**	20.64
1D-SSA	2020.78	331.35	206.63	**126.13**	157.24
Vector					
MSSA-column	400.12	33.80	18.95	**18.64**	19.31
MSSA-row	1809.04	58.67	**18.51**	18.98	20.06
1D-SSA	245.45	159.69	152.60	**148.96**	195.31

The results of investigation for different window lengths L are summarized in Table 3.4. The 24-term forecast was performed. The cells corresponding to the method with row-best forecasting accuracy are shown in bold and the overall best is in blue color.

Note that both signals are of the same structure and have rank 4. Therefore the rank of their collection is 4. One can see that 1D-SSA provides inaccurate forecasts. These numerical results are similar to the conclusions given in [23] and confirm that for recurrent and vector forecasting, the choice of window length differs for different forecasting methods. Also, vector forecasting is more accurate for the proper choice of parameters; this agrees with similar conclusions from [23, 29] for time series without trends.

Different forecasts of wine sales. Consider the time series with sales of 'Fortified wine' and 'Dry wine'. They have similar range of values and therefore we do not scale them. Let us compare different forecasts. As the structure of the time series is similar to the sum of a linear trend and sinusoids due to seasonality, the window length is chosen in view of results given in Table 3.4. The length of both time series is equal to $N = 176$. Therefore, we take $L = 118$ (approximately equal to $2N/3$) for the recurrent forecasting by rows and for vector forecasting by columns. The window length $L = 88$ (equal to $N/2$) was taken for the recurrent forecasting by columns and for vector forecasting by rows. Figure 3.9 shows that the forecasts are very similar.

3.10.2 2D-SSA

Recall that in 2D-SSA, the column space consists of spanned windows of size $L_1 \times L_2$ while the row space consists of spanned windows of size $K_1 \times K_2$, where $K_i = N_i - L_i + 1$. The trajectory matrix has size $L_1 L_2 \times K_1 K_2$. Formally, the column

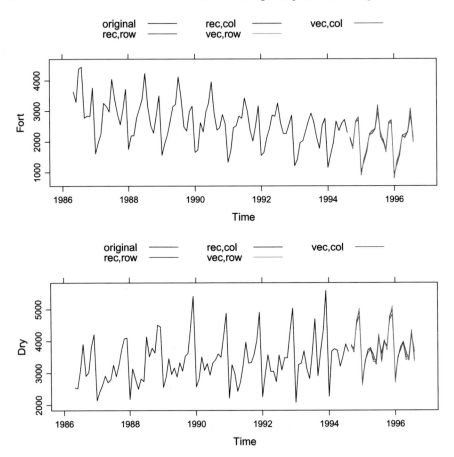

Fig. 3.9 Fortified and Dry wines: SSA forecasting

and row spaces have the same sense up to values of L_i and K_i. Let us briefly discuss the differences from the 1D case; they are mostly related to the rank and LRRs.

Ranks and separability. For 1D data, any rank-deficient infinite-length series has finite rank. For 2D-SSA, this is not so. For example, for the image with $x_{ij} = \sin i \cdot \ln(j+1)$ and any window size (L_x, L_y) such that $2 \le \min(L_x, K_x) \le \lfloor N_x/2 \rfloor$ $1 \le \min(L_y, K_y) \le \lfloor N_y/2 \rfloor$, the rank of the trajectory matrix is $2 \cdot \min(L_y, K_y)$.

Another difference from the 1D case is that the ranks of interpretable components in the multidimensional case are generally larger. For example, the SSA rank of the harmonic $x_n = A \cos(2\pi \omega n + \phi)$ is equal to 2 for $0 < \omega < 0.5$. As follows from [19, Proposition 4.7], the 2D-SSA rank of a 2D harmonic can vary from 2 to 4. For example the rank of $x_{kl} = A \cos(2\pi \omega^{(X)} k + 2\pi \omega^{(Y)} l)$ equals 2, whereas the rank of $x_{kl} = A \cos(2\pi \omega^{(X)} k) \cos(2\pi \omega^{(Y)} l)$ equals 4.

Consider a 2D array (x_{kl}) with elements $x_{kl} = y_k z_l$, where $\mathbb{Y} = (y_1, \ldots, y_{N_1})$ has rank d_1 and $\mathbb{Z} = (z_1, \ldots, z_{N_2})$ has rank d_2. It follows from [19, Theorem 2.1] that the 2D-SSA rank of this array is equal to $d_1 d_2$ and the singular values of its trajectory matrix consist of products of the singular values of $\mathbb{Y}$ and $\mathbb{Z}$. This explains why the ranks of 2D components should be expected to be large and why the singular values of 2D arrays could be very different. In the signal plus noise 2D model, the latter would almost inevitably imply the lack of strong separability.

Linear recurrence relations. The theory of nD-SSA is related to such notion as ideal of polynomials in n variables [18]. In Sect. 3.2, we described how LRRs for time series are connected with polynomials of one variable called characteristic polynomials. In a similar way, LRRs for nD objects are connected with polynomials in n variables. The existence of a unique minimal LRR in the 1D case is caused by the fact that each ideal in the ring of polynomials in one variable is principal; that is, it is generated by a single polynomial. If $n > 1$, this is not so. There is a set of generating polynomials and, therefore, a set of LRRs, which can be used e.g. for forecasting and gap filling. The set of generating polynomials is closely related to Gröbner bases. This technique is used in the case of arrays, which can be continued in all directions. Since collections of time series, where MSSA is applicable, can be continued in one direction only, the discussion above is not relevant to MSSA. This is despite the fact that from the algorithmic point of view, MSSA is a particular case of 2D SSA.

Forecasting, gap filling, parameter estimation. Forecasting for digital images is not an appropriate task except for the case when one of the dimensions is time while the other dimensions are spatial. Generally, in the nD case with $n > 1$, there is no single LRR that can perform forecasting in several directions and hence a set of LRRs (which is not uniquely defined!) performs this task. This specificity holds for the problem of subspace-based gap filling. Iterative gap filling can be performed in exactly the same manner as in the 1D case [7].

Estimation of parameters by the ESPRIT method (see Sect. 3.8.2) can be extended to the 2D case; the related 2D-modification is called 2D-ESPRIT [41]. The estimation of frequencies by 2D-ESPRIT based on the use of Hankel-block-Hankel trajectory matrices appearing in 2D-SSA, is rather popular. The R code for 2D-ESPRIT is considered in [24, Sect. 5.3.4].

Filtering. Adaptive filtering is an important application of SSA. Let us show an example, which visually illustrates the filtering ability of 2D-SSA. The image of Saturn's rings was taken in visible light with the Cassini spacecraft wide angle camera on Oct. 27, 2004, https://photojournal.jpl.nasa.gov/catalog/PIA06529 (image credit to NASA/JPL/Space Science Institute). We consider the version of this image in resolution 320×512 and note that we have removed the Saturn's moons from the image. After 2D-SSA decomposition with the window 50×50, we obtain the decomposition depicted in Fig. 3.10. The reconstruction by ET1–8 shows a general form of the rings' image, the reconstruction by ET9–30 reflects moderate changes and finally the residual shows sharp changes. The latter can help in distinguishing fine image details.

Real-life applications related to filtering abilities of 2D- and 3D-SSA can be found in [22, 43, 44].

Fig. 3.10 Saturn's rings: Original image (left,top), reconstructions by ET1–8 (right,top), ET 9–30 (left,bottom) and residuals (right, bottom)

References

1. Badeau R, David B, Richard G (2004) Selecting the modeling order for the ESPRIT high resolution method: an alternative approach. Proc IEEE ICASSP 2:1025–1028
2. Badeau R, Richard G, David B (2008) Performance of ESPRIT for estimating mixtures of complex exponentials modulated by polynomials. IEEE Trans Signal Process 56(2):492–504
3. Barkhuijsen H, de Beer R, van Ormondt D (1987) Improved algorithm for noniterative time-domain model fitting to exponentially damped magnetic resonance signals. J Magn Reson 73:553–557
4. Beckers J, Rixen M (2003) EOF calculations and data filling from incomplete oceanographic data sets. Atmos Ocean Technol 20:1839–1856
5. Bezerra LH, Bazan FSV (1998) Eigenvalue locations of generalized companion predictor matrices. SIAM J Matrix Anal & Appl 19(4):886–897
6. Bozzo E, Carniel R, Fasino D (2010) Relationship between singular spectrum analysis and Fourier analysis: theory and application to the monitoring of volcanic activity. Comput Math Appl 60(3):812–820
7. Jannis von Buttlar, Zscheischler J, Mahecha M (2014) An extended approach for spatiotemporal gapfilling: dealing with large and systematic gaps in geoscientific datasets. Nonlinear Process Geophys 21(1):203–215
8. Cadzow JA (1988) Signal enhancement: a composite property mapping algorithm. IEEE Trans Acoust 36(1):49–62
9. Chu MT, Funderlic RE, Plemmons RJ (2003) Structured low rank approximation. Linear Algebra Appl 366:157–172
10. Efron B, Tibshirani R (1986) Bootstrap methods for standard errors, confidence intervals and other measures of statistical accuracy. Stat Sci 1(1):54–75
11. Gel'fond A (1971) Calculus of finite differences. Translated from the Russian. International monographs on advanced mathematics and physics. Hindustan Publishing Corp., Delhi

12. Gillard J, Zhigljavsky A (2013) Optimization challenges in the structured low rank approximation problem. J Global Optim 57(3):733–751
13. Gillard J, Zhigljavsky A (2016) Weighted norms in subspace-based methods for time series analysis. Numer Linear Algebra Appl 23(5):947–967
14. Golyandina N (2010) On the choice of parameters in singular spectrum analysis and related subspace-based methods. Stat Interface 3(3):259–279
15. Golyandina N (2020) Particularities and commonalities of singular spectrum analysis as a method of time series analysis and signal processing. WIREs Comput Stat 12(4):e1487
16. Golyandina N, Osipov E (2007) The "Caterpillar"-SSA method for analysis of time series with missing values. J Stat Plan Inference 137(8):2642–2653
17. Golyandina N, Stepanov D (2005) SSA-based approaches to analysis and forecast of multidimensional time series. In: Proceedings of the 5th St.Petersburg Workshop on Simulation, June 26-July 2 2005. St. Petersburg State University, St. Petersburg, pp 293–298
18. Golyandina N, Usevich K (2009) An algebraic view on finite rank in 2D-SSA. In: Proceedings of the 6th St.Petersburg Workshop on Simulation, June 28-July 4, St. Petersburg, Russia, pp 308–313
19. Golyandina N, Usevich K (2010) 2D-extension of singular spectrum analysis: algorithm and elements of theory. In: Olshevsky V, Tyrtyshnikov E (eds) Matrix methods: theory, algorithms and applications. World Scientific Publishing, pp 449–473
20. Golyandina N, Zhigljavsky A (2020) Blind deconvolution of covariance matrix inverses for autoregressive processes. Linear Algebra Appl 593:188–211
21. Golyandina N, Nekrutkin V, Zhigljavsky A (2001) Analysis of time series structure: SSA and related techniques. Chapman&Hall/CRC, London
22. Golyandina N, Usevich K, Florinsky I (2007) Filtering of digital terrain models by two-dimensional singular spectrum analysis. Int J Ecol Dev 8(F07):81–94
23. Golyandina N, Korobeynikov A, Shlemov A, Usevich K (2015) Multivariate and 2D extensions of singular spectrum analysis with the Rssa package. J Stat Softw 67(2):1–78
24. Golyandina N, Korobeynikov A, Zhigljavsky A (2018) Singular spectrum analysis with R. Springer, Berlin
25. Golyandina N, Korobeynikov A, Zhigljavsky A (2018) Site-companion to the book 'Singular spectrum analysis with R'. https://ssa-with-r-book.github.io/
26. de Groen P (1996) An introduction to total least squares. Nieuw Archief voor Wiskunde 14:237–253
27. Hall MJ (1998) Combinatorial theory. Wiley, New York
28. Harris T, Yan H (2010) Filtering and frequency interpretations of singular spectrum analysis. Phys D 239:1958–1967
29. Hassani H, Mahmoudvand R (2013) Multivariate singular spectrum analysis: a general view and vector forecasting approach. Int J Energy Stat 01(01):55–83
30. Kondrashov D, Ghil M (2006) Spatio-temporal filling of missing points in geophysical data sets. Nonlinear Process Geophys 13(2):151–159
31. Kumaresan R, Tufts DW (1980) Data-adaptive principal component signal processing. In: Proceeding of the IEEE conference on decision and control, Albuquerque, pp 949–954
32. Kumaresan R, Tufts DW (1983) Estimating the angles of arrival of multiple plane waves. IEEE Trans Aerosp Electron Syst AES-19(1):134–139
33. Kung SY, Arun KS, Rao DVB (1983) State-space and singular-value decomposition-based approximation methods for the harmonic retrieval problem. J Opt Soc Amer 73(12):1799–1811
34. Markovsky I (2019) Low rank approximation: algorithms, implementation, applications (Communications and Control Engineering), 2nd edn. Springer, Berlin
35. Markovsky I, Usevich K (2014) Software for weighted structured low-rank approximation. J Comput Appl Math 256:278–292
36. Nekrutkin V (2010) Perturbation expansions of signal subspaces for long signals. Stat Interface 3:297–319

37. Oppenheim AV, Schafer RW (1975) Digital signal processing. Prentice-Hall, Upper Saddle River
38. Pakula L (1987) Asymptotic zero distribution of orthogonal polynomials in sinusoidal frequency estimation. IEEE Trans Inf Theor 33(4):569–576
39. Pepelyshev A, Zhigljavsky A (2010) Assessing the stability of long-horizon SSA forecasting. Stat Interface 3:321–327
40. Roy R, Kailath T (1989) ESPRIT: estimation of signal parameters via rotational invariance techniques. IEEE Trans Acoust 37:984–995
41. Sahnoun S, Usevich K, Comon P (2017) Multidimensional ESPRIT for damped and undamped signals: algorithm, computations, and perturbation analysis. IEEE Trans Signal Process 65(22):5897–5910
42. Schoellhamer D (2001) Singular spectrum analysis for time series with missing data. Geophys Res Lett 28(16):3187–3190
43. Shlemov A, Golyandina N, Holloway D, Spirov A (2015) Shaped 3D singular spectrum analysis for quantifying gene expression, with application to the early *Drosophila* embryo. BioMed Res Int 2015(Article ID 986436):1–18
44. Shlemov A, Golyandina N, Holloway D, Spirov A (2015) Shaped singular spectrum analysis for quantifying gene expression, with application to the early *Drosophila* embryo. BioMed Res Int 2015(Article ID 689745)
45. Stoica P, Moses R (1997) Introduction to spectral analysis. Prentice Hall, Upper Saddle River
46. Tufts DW, Kumaresan R (1982) Estimation of frequencies of multiple sinusoids: making linear prediction perform like maximum likelihood. Proc IEEE 70(9):975–989
47. Usevich K (2010) On signal and extraneous roots in Singular Spectrum Analysis. Stat Interface 3(3):281–295
48. Van Huffel S, Chen H, Decanniere C, van Hecke P (1994) Algorithm for time-domain NMR data fitting based on total least squares. J Magn Reson Ser A 110:228–237
49. Zhigljavsky A, Golyandina N, Gryaznov S (2016) Deconvolution of a discrete uniform distribution. Stat Probab Lett 118:37–44
50. Zvonarev N, Golyandina N (2017) Iterative algorithms for weighted and unweighted finite-rank time-series approximations. Stat Interface 10(1):5–18
51. Zvonarev N, Golyandina N (2018) Image space projection for low-rank signal estimation: Modified Gauss-Newton method. arXiv:1803.01419

Printed in the United States
By Bookmasters